PROTE
BUILDINGS

*How to Combat
Dry Rot, Woodworm and Damp*

Stanley A Richardson

The fabric and contents of any building, large or small, are under continuous attack. Every house owner, every custodian of an historic building, even the boat or caravan owner, needs to guard against and treat some of the ravages of weather, wear-and-tear, insect pests and fungi.

The causes and treatment for all the major problems that may afflict a house — or a cathedral or any other building — are identified, with excellent illustrations. Wet and dry rot puzzle many a householder: here are explanations and illustrations, with information on how to tackle them. Woodworm, death-watch beetle and other damaging insects can be warded off or eliminated once you know what you are doing. Paint, varnish and preservatives will do much to protect wood, stone, brick and metal, but not if applied without understanding; damp-proofing precautions are futile without adequate ventilation.

The book is the result of a lifetime's experience in caring for buildings, and is straightforward, highly readable and enlivened with humorous anecdotes — about people as well as pests! For the uncertain householder who 'must get it fixed over the weekend', or for the surveyor or architect responsible for a maintenance operation, here is information presented in a way that can be savoured as well as turned to practical account.

PROTECTING BUILDINGS

PROTECTING BUILDINGS

How to Combat Dry Rot, Woodworm and Damp

Stanley A Richardson

David & Charles
Newton Abbot · London · North Pomfret (VT) · Vancouver

ISBN 0 7153 7321 8
Library of Congress Catalog Card Number 76-54087

First published 1977
Second impression 1978

Printed in Great Britain
by Biddles Limited Guildford
for David & Charles (Publishers) Limited
Brunel House Newton Abbot Devon

Published in the United States of America
by David & Charles Inc
North Pomfret Vermont 05053 USA

Published in Canada
by Douglas David & Charles Limited
1875 Welch Street North Vancouver BC

Contents

Foreword

No man has ever been able to halt entirely the ravages of time on the buildings he constructs. The decay may be biological, the result of chemical action, or of mechanical fatigue from 'wear and tear'. It may come from 'weathering' by wind, rain, frost, snow, or less obviously, heat and ultra-violet ray. Whatever the cause, from the laying of the first brick and timber, the process begins, affecting every material used in the construction.

Paint, varnish and special preservatives will do much to protect wood, stone, brick and metal from these destructive influences, but their application must be careful, regular and, above all, discriminating. These measures are pointless without constant vigilance to detect broken and blocked rain-water goods and repair defective plumbing and drains. Equally, damp-proofing precautions and fabric maintenance are futile without adequate ventilation. Keeping a building safe and sound is impossible without attention to all these interdependent aspects of good maintenance, and expensive protection against animal and vegetable pests are a waste of money if the basic, continuous tasks are not carried out at the right time, with the right materials, in the right way.

It is for the house owner, the custodian of an historic building, even for the boat or caravan owner, indeed anyone who is burdened with responsibility for maintenance, that this book is intended. Armed with some knowledge of the causes of deterioration and a thorough knowledge of the methods and materials to combat them, much of this depressing catalogue of defects can be averted, often at a fraction of the expense which results from further delay or a botched job.

It is to be hoped that, by presenting the cause and cure of each of the major problems that afflict buildings clearly and comprehensively, *Protecting Buildings* will be as useful to the surveyor or architect responsible for a complex maintenance operation, as it should be for the unschooled householder who 'must get it fixed over the weekend'.

Chapter I

Know Your Enemies

My own interest in timber decay dates from about 1931. I had spent some years on the Gold Coast where I worked for a European company as a pharmaceutical chemist and studied entomology as a hobby. Back home again in Winchester I started a retail pharmacy, but business was very slow, life was irksome and I longed for some means of occupying my energy and initiative. The chance came in a rather unexpected way. One day, a small car stopped outside my shop and a man came in with a piece of paper on which was written a very strange 'prescription' indeed. This consisted of a mixture of paradichlorobenzene, soft soap and cedar wood oil, which he wondered if I could 'dispense'. I said I could, but told him it would take a day or so to obtain the ingredients. I asked him how much he wanted and was staggered to find that he wanted regular supplies of 50 gal lots in 5 gal drums. Obviously I could not mix this quantity of chemical in my small dispensary so I rented a nearby garage.

When the customer arrived to collect his first consignment, I asked what he planned to do with this peculiar, messy and dangerous concoction. He told me that it was to be used to kill Death-watch Beetles in Winchester Cathedral and would be applied to the timber with French vineyard sprays. I pointed out that the paradichlorobenzene was dangerous and likely to cause severe inflammation of the lungs when in vapour form, but he assured me that his lungs were hardened to the chemical and he would come to no harm. Certainly he continued to call regularly for his supplies for several months. Then, without warning, the visits stopped. Not wanting to be left with a large quantity of this highly unpleasant and unsaleable material, I made enquiries at the cathedral. I then discovered that my customer had died when treating a church in another part of the country and his death was due to the effects of the chemical.

However the cathedral architect was very anxious that someone should continue the work of eradicating death-watch from the roof timbers and eventually, when it was realised that not only was I a pharmaceutical chemist but also very interested in entomology, I was asked if I could help. Would I make a study of the Death-watch Beetle and try to evolve a formulation which would destroy the beetles but *not* the men who applied it?

There was little reliable information available, so I began by obtaining from the cathedral a piece of heavily infested oak. By carefully splitting the wood I was able to reveal the larvae working in their galleries. I kept the wood in the damp cellar of my shop and then covered the revealed galleries with a sheet of glass. In this way, I was able to study the growth rate of the larvae, the production of excrement, muscular action and methods of gnawing.

At the same time, I was able to investigate the effect of a variety of chemicals on the larvae. What basically was needed was something which was not only a contact poison but would also affect the respiratory and digestive processes of the insect. When at last I was satisfied, work began again on the timbers of the retro-choir roof of the cathedral.

Because of the presence in the formula of trichlorethylene (a powerful anaesthetic) and Rotenone (extract of derris) which had an unpleasant effect on the tongues and throats of the men applying the material, I insisted that they should wear respirators. The somewhat macabre effect of the masks worn by men wielding French vine-spraying spray lances caused a press photographer to send some rather sensational photographs to the news agencies: one of these appeared in a London daily under the heading 'Death to the Death-watch', with a colourful account of the experimental work in process at the cathedral. This unexpected publicity resulted in vast numbers of enquiries from church authorities, architects, diocesan surveyors and the owners of old properties all over Britain. Winchester's timber-decay specialists were in business.

Coincidentally, Bertram Starling, a casual friend who had become interested in my experiments, and was himself a petroleum chemist, inherited £500 unexpectedly and suggested the formation of a company to handle my formula, then known as Anobol. So, in July 1935, the company named Richardson & Starling Limited was launched. We quickly decided to employ and train men to apply the formula by methods which we had to evolve, and with equipment which we had to invent or adapt to our purpose. As unemployment was rife, we had no great difficulty in selecting a very willing team who soon made themselves expert in the difficult task of treating the complicated timber structures of church roofs, furniture and flooring. Every day was exciting as more and more work poured in. Cathedrals, abbeys, minsters, palaces, colleges and stately homes were dealt with, each job increasing our knowledge and expertise, but in 1939 B. J. Starling decided to change to the motor business and so he resigned his directorship.

Winchester Cathedral has always been regarded as the birthplace of the company, and the name Wykamol, eventually given to the main product, was derived from that of William of Wykeham, the illustrious bishop who was largely responsible for the giant restoration and modernization project carried out at the end of the fourteenth century, which gave the cathedral its fluted arcade pillars, vaulting and 'new look' from the original Norman style of the eleventh century.

The vaulting over the choir and retro-choir of the cathedral consists entirely of wood, mostly oak. The great fluted groins rest upon stone corbels built into the wall. From above—that is from the roof space between the ceiling and the roof—the vaulting falls away into deep funnel-shaped pits.

Many years after our first treatment work at the Cathedral, it was decided to repaint and regild the ornamental bosses above the choir, which necessitated the erection of enormous scaffolding to enable the team of volunteers to reach the vaulting. This provided the first opportunity to examine the underside of the vaulting at close quarters and it was noticed that the timber of the groins over each corbel was riddled with flight holes of the Death-watch Beetle. The cathedral architect asked me to inspect the remaining strength of the wood. This I could only assess by drilling with an auger to discover how deeply the beetles had penetrated. The 1in auger bit into the outer crust to a depth of only about half an inch and then met no resistance at all. When it was withdrawn from the wood, bore dust

11

immediately poured from the hole like water from a tap, fell
through the scaffolding and began to spread over the floor 80ft
below. Like the famous little Dutch boy at the dyke, I pushed
my finger into the hole and waited for someone to come along
so that I could raise the alarm. Fortunately it was not long
before the head of one of the builder's men appeared at the top
of the ladder (which extended up to the platform on which I
was standing). The situation was explained to him and he
hurried away, returning in record time with an empty five-gallon
preservative drum. I withdrew my finger and the tiny pellets
streamed into the container. In a very few minutes, the drum
was filled and I returned to my dyke-act until it could be
replaced with another. Altogether, ten 5gal drums of bore dust
were collected from the vaulting groins of the ceiling over the
choir.

This indicated that the great moulded groins were practically
hollow and in a precarious state. A considerable engineering
feat had to be performed so that the weight of the ceiling could
be supported while the lower part of each groin was cut away
and an exactly matching piece inserted in its place. The timber
had suffered so severely because moisture from the leaks in the
roof gravitated to the bottom of each groin where the moisture
set up fungal decay. This in its turn encouraged the activity of
the Death-watch Beetle.

During the early life of my company another interesting
treatment was undertaken at St Mary's Hospital, Chichester.
This was originally a monastery, erected in the thirteenth
century, but after the Dissolution the building was converted
into almshouses. These were in the form of roofless cabins or
tiny bungalows built within the walls and under the canopy of
the great roof. From the outside, the building looks like a
church with a huge tiled roof set on low walls. Being built on
marshy ground the foundations were far from stable and, as was
discovered in 1939 when the roof was examined, the roof
timbers were seriously Death-watch infested. Even worse, the
wall on the south side was being forced outwards by the weight
of the roof timbers; nearly all the dowels holding the joints
together had perished, and were pulling out of the mortices;
leaving, in many cases, less than a quarter of the tenon in its
socket. The architect sent for an eminent consultant who

accepted that the roof was unsafe and suggested that a false roof be built over the existing one, from which the original roof could be suspended by iron rods.

The cost was prohibitive and it was decided to handle the beetle treatment and leave the structural work until funds could be raised. The almshouses were all occupied by elderly ladies who stayed in their tiny houses despite the dust and noise. One day the rubber hose connecting the spray nozzle to a pneumatic spray burst and the contents of the cylinder poured into the sitting room of one of the houses. My foreman discovered to his horror that the brown Wykamol had swamped an entire set of the *Encyclopaedia Britannica*. The elderly occupant of the house was out shopping and someone had to break the news to her. I tried to think what on earth I could say to the poor old soul who must obviously cherish these great tomes and their link with her past. I waited for her to arrive and then it took some time before I could make her understand what had happened.

'Do you mean,' she demanded, 'that you have ruined my late husband's set of *Encyclopaedia Britannica?*'

'Yes,' I admitted in a whisper.

'Oh thank God,' she answered, 'I've always regretted keeping those heavy things. Now I shall never have to dust them again.'

Hitler's bombers unwittingly performed a great service to St Mary's during the latter part of the war when Chichester Cathedral was ringed with bombs and one fell on the south side of St Mary's. The blast shook off all the tiles but it also pushed the south wall upright and the loose timbers went back into place. I suggested that buttresses to hold up the south wall could settle the roof problem and the cost be included in the war-damage claim. Buttresses were inserted, all dowels removed and the building completely restored at comparatively little cost.

In the early days of Richardson & Starling Limited, cash was very tight and the men and I had to exercise great care not to involve ourselves in unnecessary expense. We were called in to treat a large church at Kingsclere. The roof timbers were fully exposed in the nave which was exceptionally high; scaffolding would obviously be needed. I set off to Newbury to find out how much it was going to cost and was quoted a figure which I, and my foreman, considered exorbitant. I went back to Winchester and the men continued, as I thought, to work on the more accessible timbers while I tried to get other quotations for scaffolding.

A day or so later I had a telephone call from a highly agitated vicar. He told me that he had found one of my men straddled across one of the small ties in the apex of the roof, applying spray treatment with no ladder in sight. The vicar could not understand how the man had ever reached this position, so he watched until he was ready to come down again. Then the other men, below, produced two ladders tied together. Even this did not reach the collar tie where their companion was marooned, so they lifted the ladder and the three of them held it several feet from the ground; and from his hazardous position the operative climbed on to the swaying ladder and lowered himself to the ground. On hearing this, I dashed over to the church to stop the foolhardy exercise but by the time I arrived all the work had been completed without the cost of scaffolding.

On another occasion I was myself guilty of a certain foolhardiness. A church steeple had to be treated very urgently for damage by Death-watch Beetle and I dealt with this myself. Having arrived, I realised that I had forgotten to bring a gas mask. Altogether there were three ladders which led up to the apex of the steeple and having climbed to the top of the third, I decided that the safest plan was to start spraying from the bottom of the steeple and work upwards because the vapour, being heavier than air, would stay below, and I would be breathing reasonably fresh air where I was working. All was well until at last I had to treat the apex of the steeple right over my head and of course the vapour then surrounded me. I continued to spray for several minutes and then quietly passed out.

When I woke again, I was at the foot of the ladder — miraculously unharmed, but considerably dazed. This was due to the trichlorethylene in the vapour. I staggered out to the car, drove home to Winchester and once in the house sat down and fell asleep. By the following day I had recovered but found I had left all my equipment behind. Unfortunately it began to snow and continued so heavily that I was unable for some time to go back to the church to collect my materials — then to find that the spray had been stolen.

Another large church, in Sussex, had wind braces running

14

from the external ribs of the steeple and meeting at a diagonal angle on the kingpost which ran vertically down the centre. During the course of treatment one of our men caught his leg in the angle of one of the wind braces and was suspended in agony, having put out the cartilage as he tried to free himself. No one could reach him and the Fire Brigade had to be summoned to rescue him.

An appeal for funds at this church was so successful that it covered far more than the cost of the Death-watch treatment and the vicar therefore decided to replace the hand-operated organ with one operated by electricity. To get the cable from the mains to the organ they had to dig a trench right through the sanctuary and in doing so cut through the grave of a woman who had died about 1650. They then had to remove what remained of the coffin and lift out the bones in order to run the cable through the grave, the bones and coffin afterwards being replaced and covered with a large flagstone lid. This aroused no comment because no one knew it had been done, but about a week later one of the workmen involved was taken ill and his doctor, very puzzled, realised that he seemed to have the symptoms of bubonic plague. The man was asked if he had been in contact with any foreign people or ancient graves and had to admit that he had removed and replaced the bones from the grave in the chancel; he was rushed to the isolation hospital and successfully treated. The vicar, thinking the story might be good publicity for his church funds, talked to the local press: this turned out to be a big mistake, as it then became clear that he had undertaken improvements in the church without first obtaining a faculty, a crime in ecclesiastical circles. The poor vicar ended by appearing before a consistory court.

Although much of our work in those early days was treatment of cathedrals, stately homes and churches, there was also a good selection of less imposing buildings. On one occasion I was asked by a well-known local architect to examine the condition of the timbers of a large store in the district while he himself checked the general structure. The architect had gone ahead of me and was about two rooms away when I decided to go out of a skylight and on to the roof to draw a general plan of the building. I placed a ladder against the purlin and climbed up, a height of only about 12 feet, but when I tried to push the skylight up to get on to the roof I found it was stuck with paint. I gave a great heave to move it and the ladder slipped away from beneath me. I managed to grab the purlin and hang suspended from it, not daring to drop down on to the floor from that height as the boards were so weakened by woodworm. I called to the architect who merely answered that he would be along in a minute. I shouted and shouted and at last he arrived, peering round the room to discover where I was. When, eventually, he spotted me hanging from the purlin, he demanded 'How the hell did you get up there?'

I said that I would explain later — the immediate essential was to replace the ladder for me! It was not easy. The architect suffered from heart trouble and it was a struggle for him to get the ladder up to the purlin again. I then had to swing myself round, hanging with one hand and then the other until I could hook my leg through the underside of the ladder and let go of the purlin, staying on the underside of the ladder until I had at last reached the comparative safety of the worm-eaten floor.

On another occasion I was inspecting the roof of a local church with the son of the architect mentioned above, and was following him along the collar beams (these tie the rafters together to form the triangular apex of the roof). Suddenly there was a crack and he dropped; the collar on which he had been standing had collapsed. Fortunately he was able to grab the collar beams on each side and was left suspended by the armpits. This probably saved his life for otherwise he would have crashed through the lath and plaster ceiling to the church floor some 35 feet below.

Such are some of the physical hazards of surveying ancient buildings but other hazards are encountered. For example, just after the 1939-45 war we had applied treatment to the roof timbers of a church. The following year we were summoned to investigate what was alleged to be a complete failure of the treatment: it was reported that many beetles were still being picked up from the church floor. We could find no evidence whatever of recent beetle emergence from the roof timbers we had treated, but there were certainly beetles crawling about the floor. Much of the floor showed signs of having been renewed; the boards were bright new pine with no signs of infestation and the rector assured me that all the old flooring, including hidden timber, had been completely removed only a few months earlier; the beetles could not possibly be emerging from this new wood. Since this could not be disputed we offered to carry out a second treatment to the roof free of charge, using labour we badly needed elsewhere. But this did not cure the trouble and beetles still wandered about the bright new flooring.

I visited the church a number of times to try to solve the mystery and one day, when I was sitting in a pew trying to think of something which might have been missed, an old man came towards me, apparently the sexton. I told him how puzzled I was and he said,

'Ah, but it b'ain't the roof, it be them joices under the floor.'

'But the parson said all flooring timbers were renewed.'

"E don't know what 'e's talking about,' answered the old fellow. 'Jist you take up a board and you'll see!"

I fetched a bolster and hammer from the car and removed a short board. Sure enough, the old oak joists underneath were heavily infested with wet rot and Death-watch Beetles. Needless to say the rector offered no compensation for the unnecessary trouble and expense caused by his misleading and irresponsible statements about the floor.

Although such incidents are, fortunately, not daily occurrences, they are nevertheless typical of the variety of problems and hazards which I have faced since my first encounter with the scurrying battalions of decay, nearly forty years ago. Each problem has had different facets and demanded different solutions, and, as that story shows, it is not always just the enemy within that makes the task a difficult one.

Chapter II

Deterioration:
a Summary of Causes

It is noticeable, wherever one goes, that the buildings which have survived the passage of several centuries are generally built of stone. In most countries, the stone buildings were those of considerable public, religious or commercial importance or those erected by the very wealthy for domestic purposes. The dwellings, shops, and farms occupied or used by the common people in ancient times were generally constructed of wood. Comparatively few of the wooden buildings built prior to the sixteenth century have survived. Those which still exist are, invariably, found to be framed with durable hardwood and in well-drained and airy situations. In England we are fortunate to have a considerable number of timber-framed buildings still surviving, but very few of the survivors date back beyond the sixteenth century, whereas many stone cathedrals and churches were built before the twelfth century, and some have been standing since Saxon days.

Many timber buildings existed before the seventeenth century, but there is hardly a town in England that cannot produce from its archives records of devastating fires which greatly reduced their numbers. Even so, it is doubtful if as much destruction was caused by fires as by biological deterioration. Fungi and insects have been responsible for the destruction and disappearance of far more buildings than fires, wars or civil disturbances.

Our awareness of the susceptibility of wood to decay seems to have existed almost as long as we have lived under shelters of our own making. Constructional timber which has been deliberately charred, pickled in brine, immersed in natural bituminous deposits, soaked in oil or in other ways protected against decay has been discovered by archeologists and many references to treatments designed to protect timber are found in some of the earliest and most exalted historical records. God, for example, advised Noah in Genesis to make an ark of gopher wood, and 'pitch it within and without with pitch'. Pitch in this case was probably the natural bituminous deposits found in so many parts of the Holy Land.

In Britain, however, little evidence exists of designed preservation of building timber until comparatively recent times. The medieval builders relied on the durable qualities of oak to provide a reasonable 'life' to their wooden structures. The method of construction also contributed to the prolonged life of many ancient buildings. Very few were provided with rain-water disposal systems but, by generous overhang of the roof and a further overhang of the upper storeys so that the ground floor was well tucked back from splash, the walls were kept fairly dry. Most of the buildings which succumbed to the effects of biological deterioration were those which became neglected as a result of the terrible poverty which occurred during the Middle Ages or were deserted in the wake of the Black Death.

Gradually the wool trade made England more prosperous and a very large proportion of the surviving medieval houses and churches serve as a memorial to the prosperous wool merchants who were directly or indirectly responsible for their erection.

There still survive a few twelfth, thirteenth and fourteenth century domestic dwellings, some with the roof timbers in a remarkable state of preservation. The lack of insect and fungal deterioration of this roof timber is due to the common practice of having a fireplace in the centre of the hall with an aperture in the apex of the roof for the smoke from the wood fires. The gases in the smoke condensed on the cool roof timbers depositing a crude form of wood creosote containing phenols and other preservative chemicals.

In the early sixteenth century, the Italians, Germans and Flemings who sought refuge from persecution in their own countries, began to influence the design of English architecture, particularly the ornamentations of wood and stone.

This elaboration and ornamentation, with carved finials, barge boards, brackets, corbels, cornices and the like increased the hazards of deterioration. Every excrescence enlarged the area on which rainwater could fall and accumulate and the maintenance of buildings built since the sixteenth century became increasingly important and more difficult.

The half-timbered buildings which became so popular in the late sixteenth and early seventeenth centuries were, and still are, some of the most difficult to maintain in good condition. The original infilling of wattle and daub (usually clay, cow dung and chopped straw spread over a lathing of willow, hazel or split chestnut) was both insulating and durable. Although it absorbed water, it quickly released it and this peculiar mixture was extremely flexible and adhesive. The introduction of cement created complications for, where repairs were carried out with sawn laths and a cement mixture, the panels were brittle and generally shrank away from the timber framing to cause draughts and water leaks. Because of these problems, many of the half-timbered houses were later battened all over and either weather-boarded or hung with tiles which spoiled the appearance and hid the framing timbers in unventilated spaces. When water found access to this timber, fungal spores germinated and the process of decay was commenced.

The introduction of floor levels, particularly the covering of the original earth ground floor with timber flooring, together with plaster or wooden suspended ceilings and the internal panelling of walls, all created stagnant areas where dampness of the atmosphere could pass saturation point and conditions favourable to fungal spore germination were created. The dry rot fungus, *Merulius lacrymans*, would have been exceedingly rare in the period when houses were single-storied, and floors just beaten earth or clay and cow-dung.

Even those in the great halls and castles were of stone flags simply covered with rushes or straw.

One common method of building in Hampshire was to erect two fences of boards, the fences placed parallel and about two feet or more apart, separated with lengths of wood across the space. These fences served in the manner of modern concrete shuttering and the gap between them was filled with moist and well rammed-down chalk. This simple material set like a soft stone, providing good insulation, rarely became damp inside and, if well-protected by an overhanging roof, would last a very long time. Such walls are always associated with thatched houses, many of which have survived for hundreds of years.

Only when the roofing thatch became seriously deteriorated, generally as the result of nesting birds or rats, and rain was allowed to fall directly on to such a wall, would it start to pulverize and break up. Once moisture penetrated, the first hard frost would cause the chalk to disintegrate completely.

There are several large houses in Hampshire with walls of chalk which still survive all the hazards to which such buildings are subject although many have failed because the owners, wishing to modernise the appearance of their house, had the external faces of the walls covered with cement rendering or rough cast with gravel, crushed stone or pebbles. In many instances, this new surface was decorated with paint. This impermeable coating trapped condensed moisture in the walls by preventing evaporation from the external faces. This trapped moisture caused fungi to develop in the spacers or lengths of wood which originally separated the shuttering and were always left in the chalk walls. If this fungus was *Merulius lacrymans,* it quickly spread through the chalk to the carcassing timbers of floor and roof and, unless the owner was prepared to partially demolish and rebuild, the house eventually disintegrated.

Even when brick-built walls became popular in the sixteenth, seventeenth and eighteenth centuries, it was customary to build into the brickwork at regular levels and in every wall, lengths of wood called 'bonding timbers'. They served the double purpose of tying the masonry together and acting as fixings for panelling, tapestry or other internal decorations. Neglect of rainwater disposal or any other reason for persistent dampness led to the decay of these timbers, usually with devastating consequences. Other builders continued to reveal the framing timbers but replaced the wattle and daub with brick in-filling, often ornamental and many in herring-bone pattern. The problem was to provide a satisfactory joint between the wood and the masonry and with varying expansion and contraction due to heat and water, few houses remained draughtproof or waterproof for long.

To the student of deterioration in ancient buildings, the incursion of moisture through walls, whether of timber, stone, brick, chalk or other material, and the capillary rise of dampness from moisture-absorbent foundations are obvious causes of fungal development, insect activity and other forms of deterioration in the building fabric. Timber in direct contact with damp masonry, soil or concrete will rarely resist the attacks of fungi and timber-destroying insects for long. Less obvious is the slow but steady deterioration which occurs in the roof and flooring timbers where no apparent water leaks have occurred and where the timber is well away from the damp walls. Even the framing of internal stud partitions on the upper floors is frequently found to be seriously damaged by organisms which require a relatively high moisture content to flourish. When investigated, this moisture is found to be due to condensation. In domestic buildings, considerable quantities of moisture vapour released from kitchens, laundries and even from the living rooms rose to the upper floors which were rarely heated in those days. This moisture-laden air, rising on thermal currents from the warmed rooms below, precipated water on to cold plaster and wood in the very centre of the building. With every window closed and every crack and cranny sealed against draughts, the condensing moisture accumulated in the bedrooms and attics to produce conditions favourable to the development of slow-acting fungi, particularly fungi imperfecti. Wood attacked by such fungi becomes susceptible to wood-devouring insect attack and in such conditions the Death-watch Beetle flourishes. Few houses built in the Middle Ages or even much more recently could resist the attacks of this or other wood-borers.

The remains of ancient buildings more often then not, comprise only the stone or brick walls, arcades and towers. All the structural and decorative timber has long since been destroyed, usually by biological organisms. Yet even the stone and brickwork usually show signs of deterioration which varies in extent according to the type of stone and the degree of weathering and other forms of erosion to which it is exposed.

Acids, generally, play an important part in the deterioration of buildings. Industrial pollution of the atmosphere with acidic gases carried by rain-water or fog is a very evident cause of deterioration. Sulphur dioxide is probably the most common as it is one of the gases produced by burning coal and coke. It forms sulphurous acid when dissolved in water and causes serious damage, particularly to limestone structures, lime mortar and anything containing calcium carbonate, oxide or hydroxide. Even bricks, in which tiny sea shells or pieces of shell have been heated and oxidised to form calcium oxide, will react to absorbed sulphur dioxide to form calcium sulphate which, by expansion will cause the bricks to spall and split and develop objectionable efflorescence.

But acids produced naturally are the most serious cause of deterioration as, in many instances, they are difficult to detect and control. Carbon dioxide dissolved in water to form carbonic acid is without doubt the most serious of these acids for although only a relatively small quantity of the gas exists in the atmosphere, rainwater and mist extract and distribute it over every exposed surface. If the surface is porous, penetration of the water or water vapour carrying the acid will produce an ever-increasing degree of deterioration.

Limestone, when quarried and used for constructional purposes, may for a number of years remain insoluble in water and apparently immune to the effects of the weather. But gradually, carbon dioxide is conveyed into the stone by moisture and water vapour to form soluble bicarbonates in the calcareous substrate. As the water-soluble bicarbonates are leached out of the stone by rainwater or excessive condensation, minute tunnels and cavities are formed and the stone becomes progressively more porous. Penetrating water will not only accelerate the action of the carbonic acid, but the stone is now able to act as host to algal growths. Green stains and slime begin to appear on the damp stone and, if the stone dries out, turn black and form nutrient on which other vegetable life such as lichens, mosses and advanced plant life can grow.

These vegetable growths produce their own quota of acids which, sometimes, have devastating effects on old buildings. Where acids released by lichens and mosses have been conveyed down a pitched roof by rain water, it is fairly common to find that the lead has become exceedingly thin or cut right through by the dissolving effect of the organic acid. Once the lead has been penetrated and rain water can get through, the increase in the process of deterioration is usually quite sensational; dampness inside the building results in fungal and insect activity which can produce rapid devastation.

The most common fungus found attacking timber in buildings, both ancient and modern, is *Coniophera cerebella* known by the common name of cellar fungus. The cellar fungus, often seen on dead wood in forests, does attack timber in damp cellars but it will with equal vigour attack flooring, studdings, joinery and even roof timbers if the environmental conditions are favourable. The most destructive fungus is *Merulius lacrymans* which bears the misleading common name of dry rot. Even Benjamin Johnson in 1803 remarked that it was a mistake to call this growth dry rot, as it is encouraged only by dampness. The name probably dates back to the time when decay of timber was not attributed to fungal growths but rather the reverse; that the growths were produced by the rotted wood. As timber which has been attacked by dry rot eventually appears very dry and the cubed remains are brittle and powdery, it was no doubt assumed that this form of decay was due to excessively dry conditions. In fact, the fungus requires damp but not very wet conditions for the germination of its spores and will germinate in circumstances where the ambient atmosphere is static and holding moisture at or near to saturation point.

It develops most rapidly at about 20°C and the British Isles afford it ideal conditions through much of the year. Dry rot is never found out-of-doors in Britain, no doubt because of the rather exacting conditions it requires for its development. But it occurs in buildings of all ages throughout Europe where neglect or faults have created suitably damp conditions. Both cellar fungus and dry rot will attack softwood more readily than the hardwood timbers of which ancient buildings were constructed. However, given favourable conditions, both will flourish on hardwoods although somewhat more slowly than on coniferous woods.

Dry rot often occurs after the wood has already been subject to attack by cellar fungus. Once established on damp wood, the hyphae of *Merulius lacrymans* will penetrate masonry, and conducting strands, five or even more millimetres thick, will pass through bricks, stone and coarse concrete. Along these strands or rhizomorphs is passed the moisture derived from the metabolism of the cellulose carbohydrate. Acid is released by the hyphae and so the fungus prepares the substrate in a perfectly dry environment by making it and the ambient atmosphere damp, acid and suitable for assimilation.

That so many ancient buildings have survived the ravages of time and the effects of so many different forms of biological deterioration is quite remarkable, particularly in view of the damp temperate conditions which generally prevail through the year in Britain. If very hot and dry conditions occur for only a few weeks, they can exert a marked degree of control over the development of fungi. So, in general fungal deterioration of buildings in Britain is more rapid than in most places on the European mainland.

To compensate for this, certain insects of Europe which do relatively little harm in Britain, attack timber in buildings on the continent. In fact the termite, *Reticulitermes lucifugus*, which is a common and a serious pest in Europe south of Paris, does not exist in the British Isles. The other serious pest on the continent is the House Longhorn Beetle, *Hylotrupes bajulus*, the activities of which are confined to a relatively small area of England south-west of London.

In Britain, the common range of insects which attack timber in buildings are the two anobiids, *Anobium punctatum* and *Xestobium rufovillosum;* two Lyctids: *Lyctus linearis* and *Lyctus brunneus* and the wood weevils, *Euophryum confine* and *Pentarthrum huttoni.* The anobiids are undoubtedly the most serious of the insect pests: the lyctids only attack the sapwood of hardwoods during the period of seasoning and rarely after ten years from felling, and the weevils require the wood to be already decayed by fungi.

The most common insect found attacking building timber in Britain and in many other parts of the world is the Common Furniture Beetle, *Anobium punctatum,* often referred to as 'woodworm'. The latter name was no doubt given to it by those who, in the past, had found the little white larva in infested wood but had not connected it with the tiny brown beetle which appeared in the spring and early summer. It attacks the sapwood of all the species of indigenous timber, both hardwoods and softwoods, used for building purposes in Britain.

A fungus commonly found attacking timber in ancient build ings but rarely in modern structures is *Phellinus megaloporus.* This usually germinates and develops in deep cracks or 'shakes' in persistently damp hardwood beams. Such fungal growths encourage the activity of the Death-watch Beetle, *Xestobium rufovillosum,* which favours hardwood softened by fungal activity for the purpose of ovipositing.

The name 'death-watch' is said to have arisen as the result of its mating tick being heard by watchers sitting by the bedsides of dying people. As few people recovered from serious illnesses or injuries in those days, the ticking heard by those keeping vigil became associated with death and, when heard, was thought to signify that 'death was about to cross the threshold'. This tapping is, of course, the mating display of the insect.

Eggs of the Death-watch Beetle are usually laid in cracks, crevices or the open ends of vessels. In such places, they are protected from physical damage, and when the larvae emerge they are in a position to enter the wood easily. Having established themselves, they may stay in the larval state, tunnelling through the timber for as long as ten years before pupating and emerging as adult beetles to continue their destructive line. During their larval stage, they will gnaw a gallery several feet long through heartwood as well as sapwood, provided it has hyphal threads running through it. There is no doubt that fungus forms an important constituent of the diet of the Death-watch Beetle and the period required to complete the life cycle is greatly influenced by the amount of fungus present.

The question is often asked, 'Why are these destructive pests inflicted on mankind to destroy his works?' These 'pests' together with many other organisms and creatures similarly regarded, perform the invaluable duty of scavengers. Their purpose is to destroy and consume dead trees and provide room for fresh trees to grow. A tree in a vigorous state of growth is rarely attacked by wood-destroying insects, but if the tree is weakened by age, weather conditions or physical damage, bark borers will initiate the destructive processes by introducing fungal spores, sometimes with rapid and devastating effect as in the case of Dutch elm disease. Other fungi can enter through the roots and cause decay in the heartwood, a form of tree disease known as 'butt rot' which has led to the destruction of many fine beech trees.

Because dead wood in the forest so quickly becomes infested with insects and fungi, it is very unwise to utilise the timber from a dead tree, no matter how sound it may appear. Timber from a perfectly healthy tree which has been recently felled by man is generally free from any kind of defect or damage caused by nature's scavengers, but a tree which has 'died' by any means other than the saw or axe should certainly not be used for building timber. This warning applies equally to trees artificially felled and then allowed to lie in the forest for several months. It is not unusual for the Wood Wasp to perforate the bark of softwood trees and lay eggs in the galleries it creates. Trees left lying on damp ground for even a week or so may become infected with the spores of sap-stain fungi which may cause serious discoloration of the wood.

It is also unwise to install into a building any timber to which bark is adhering as the outer sapwood to which the bark is attached is particularly vulnerable to insect attack. Even the bark attracts certain types of wood-boring insects but the damage they cause is rarely of economic importance. Nevertheless, two of these insects deserve mention because they are often mistaken for some which do cause serious damage. As a result, a great deal of money has, in the past, been spent on treatment which was quite unnecessary.

The first is the anobiid, *Ernobius mollis*. Slightly larger than the Furniture Beetle and slightly smaller than the Death-watch Beetle, it is often mistaken for one or the other. Its shape is similar to theirs and its reddish-brown colour provokes further confusion. It only attacks softwoods on which the bark is adhering. It cannot attack barked timber (timber from which the bark has been removed) because it lays its eggs in dry bark. The larvae which emerge from these eggs tunnel in the bark and, it seems, almost accidentally penetrate the outer sapwood. They never penetrate for more than a very short distance into the wood, usually less than half an inch. The damage they do has little or no significance but, on occasions, has caused a great deal of anxiety to timber merchants, builders, architects and their clients. Unfortunately it is not unusual to find building timber with some bark present, generally down one edge or face. After such timbers have been built into a house, *Ernobius* are seen to emerge, bore dust falls from the wood, flight holes appear and this then gives the impression of serious infestation.

Sometimes the timber is acting as studwork and fixing for hardboard and other linings. The beetles will often bore through these linings, leaving holes which give rise to all sorts of erroneous ideas of the susceptibility of certain fabricated boards to woodworm. If all the bark is removed *Ernobius mollis* will never occur, and if it is present, it will certainly go once the bark is removed. If the infected wood is behind finished surfaces of panelling, wallboards or even plywood, careful injection of insecticide into each flight hole as it occurs, together with a little patience, will usually produce the desired effect without disturbance of the surface or the expenditure of more then a few shillings in insecticide. Injecting into the flight holes after the insects have flown may seem futile, but the object is to introduce the chemical to the wood behind the panelling and thus reach any other larvae or insects in close proximity to the hole. Most proprietary insecticides are designed to penetrate and spread, so careful injection into one hole will often effectively treat several square inches of porous bark and sapwood.

The other bark borer which may cause confusion and on numerous occasions has been mistaken for the very destructive House Longhorn Beetle is the Longhorn which favours hardwoods, *Phymatodes testaceus.* This beetle, like *Ernobius mollis,* should be known to timber merchants, builders, architects and surveyors because it attacks seasoned timber, though generally only hardwoods and then only when the bark is attached. Damage caused by the insect is often seen in the roofs of ancient buildings and is mistaken for serious infestation. It carefully examined, it will be found that the damage is confined to sapwood which has bark still adhering to it or from which the bark has fallen away. If the wood is cut and the galleries are revealed, it will become apparent even to the untrained and non-technical eye that the borings are old and are confined to wood which obviously contributes little to the strength of the beam or structure. The holes in the surface of the wood are usually oval, and about twice the size of those of the Death-watch Beetle, which may well be present in such timber. The Death-watch Beetle holes are about $1/8$ in across and circular; those of the Longhorn are $1/4$–$1/3$ in in length, and slightly narrower across the centre. The adult insects are rarely found in houses, but are by no means uncommon in timber yards, particularly in home-grown hardwoods such as oak.

Unfortunately, the forest insects often extend their scavenging activities to timber which man has gone to great trouble to prepare, cut and shape for his own use. To these insects and, indeed, to wood-destroying fungi, timber is just dead material and, if left damp and unprotected as it is in the forests, they set about it with all the vigour they can muster, each helping the other to achieve its individual design.

Death-watch Beetles, which so often infest the timbers of old buildings — churches in particular — would not lay their eggs on dry, well-kept timber. They only attack timber in which dampness, at some time or other, has caused fungi to develop. More often than not, the fungus is undetectable with the naked eye, but microscopic examination will invariably reveal its presence. The dampness may not be due to direct contact with water but to condensation of moisture from a warm, damp atmosphere on the colder timber. In churches, for instance, spasmodic heating causes moisture to evaporate from the damp walls and floors, rise into the roof space and condense on the roofing timbers. Every time the boilers are lit and the heating system warms up, fresh supplies of moisture are carried up on the rising currents of air, and help to nourish and stimulate the destructive organisms in the roof.

Dry timber is safe against most forms of decay and deterioration. Even the ubiquitous woodworm, or Common Furniture Beetle, will rarely attack wood with a moisture content below 12 per cent. Fungal decay organisms such as the so-called wet rots and dry rot itself will never occur either on wood with a similarly low moisture content.

So the finest defence against timber-destroying insects and fungi is warmth and ventilation which, together, reduce dangerous humidity and dampness. Alternatively, even damp timber can be kept free from decay by timber preservatives containing long-lasting toxic chemicals which destroy insects and fungi and inhibit egg-laying and spore germination.

Chapter III

Masonry and Dampness

There are, of course, many different types of stone used for building. Very broadly, they may be classed as sandstone, limestone, flint, granite or igneous and metamorphic rock, but among them are infinite variations of composition, texture and physical properites. Although stone in general is the most durable of the materials used in the construction of ancient buildings and shrines, it will eventually break down and disintegrate. Left in its natural bed, only a very tiny proportion of the whole is exposed to variations of temperature, moisture, abrasion and so on, but, when used in buildings, a higher proportion becomes exposed to these variables and the extra hazards they create. Alternate drying and wetting, sometimes with freezing, will have an erosive effect. The mechanical action of wind and rain, the chemical action of acids conveyed into the stone by the rain, the growth of algae, lichens, mosses and higher plant life, together with the calamitous effect of iron dowels, cramp and fixings, cause deterioration and disintegration.

Sandstones consisting mainly of quartz grains held in a matrix of silica, calcite, magnesium carbonate or some other binder are, generally, as durable as the nature of this binder permits. Siliceous sandstones are extremely durable, but the others vary from good to extremely bad. In argillaceous sandstones, the grains are held together by clay and, as such, are probably the least durable. However, deterioration is mostly dependent upon dampness and when the majority of the building materials are kept dry, it is slow. In damp or alternatively wet and dry conditions, it can be greatly accelerated.

Another and perhaps the most serious cause of damage to stone is the crystallisation of salts contained in solution in the masonry. These salts may be absorbed with moisture from the soil, or they may be deposited by spray from the sea or waterfalls or even from street water splashed by passing vehicles. Atmospheric pollution by harmful gases may produce acids which, when dissolved in rain or mist, react with the stone itself to form such salts. However these deposits occur, they may remain comparatively harmless whilst in a weak solution but, if the stone dries, the solution concentrates and crystallisation takes place, damage quickly occurs. As the drying-out of the stone generally occurs more frequently and for longer periods on the inside faces of walls, the disintegration will, generally, be most apparent where the walls appear to be most protected. In one case, sand was collected from the seashore for mortar to carry out repair works and repointing. The salt from this sand, being deliquescent, gradually migrated through sandstone mullions and window surrounds. Once inside, it began to crystallize and effloresce, causing serious pulverization to the interior of the building.

Flints, of which many buildings were constructed in medieval times, although exceedingly durable in themselves, had to be bound together and set in clay or lime mortar. The life of the building thus depended largely on the constituents of the mortar and the knowledge and skill of the builder.

Many methods have been tried for arresting the decay of stone masonry but, at present, only those which keep the stone dry are at all satisfactory. One of the oldest — and still used on limestone and flint — is to coat the wall regularly with a solution of lime and water. For the preparation of this 'lime-water', the water must be soft, preferably rainwater collected in wooden casks. Lump quicklime is then slaked and allowed to settle. The clear liquid decanted from the surface of the lump will contain a saturated solution of the lime which is applied liberally to the surfaces to be treated. This solution of calcium hydroxide quickly takes up carbon dioxide from the atmosphere to form the carbonate which fills all minute cracks, pores and interstices. It still remains sufficiently porous to permit water vapour in the wall to evaporate.

Indeed, any application which forms an impervious skin on the surface of the stone is most unsatisfactory, as it will prevent the evaporation of trapped moisture from the external face of the wall. This trapped moisture will accumulate until it appears on the inner face, and the conditions of dampness it was aimed to cure will be greatly intensified. This becomes particularly bad in winter when windows are closed and ventilation reduced. In such cases, the walls frequently reach over-saturation point and condensed moisture then runs down the inner faces.

During the last decade, a great deal of work has been done on the production of silicone resins for use as water repellents on masonry. Silicones are high-polymeric, silicon-rich organic compounds. A very wide variety of them now have industrial applications, ranging from lubricants, insulators, synthetic rubbers to the silicone resins which have hydrophobic properties and closely resemble petroleum waxes. At first they were not all satisfactory for use on limestones, but modification of the molecular structure has produced highly satisfactory results. Silicones now available provide good and prolonged protection for most forms of masonry, including brickwork and clay tiles.

Alcoholic solutions of siliconester are occasionally used to bind friable stonework by hydrolization and deposition of silica. Unfortunately, the penetration of such a solution is generally very limited, and only a hard crust is formed which can be blown off by crystallisation due to freezing or soluble salts trapped behind the treated area. If the wall is very dry when treated and an organic solvent solution of silicone resin is applied to the treated spot as well as a wide area surrounding it before moisture builds up again, the results can be reasonably satisfactory.

Masonry Preservation

For centuries, men have been trying to neutralize the destructive forces of nature and, in many spheres of human endeavour, have achieved considerable success but, until recently, very little progress had been made in the preservation of stone.

There are relatively few people with sufficient knowledge of stones, the appropriate scientific background and the experience to be regarded as true experts on stone preservation. Whoever attempts to preserve stone should have a good knowledge of stone in general and an intimate knowledge of the stone it is intended to treat in particular, together with a basic knowledge of the practical and chemical problems involved. Of the few

Plate 1 Fruiting bodies on an external wall in an unventilated area

Plate 2 Typical indication of dry rot behind matchboard panelling

Plate 3 Dry rot in skirting boards

Plate 4 Merulius fruiting bodies – note the electric socket where a short circuit could cause danger from fire

expert sources of advice, the Penarth Laboratories at Otter-bourne in Hampshire are probably the most knowledgeable.

Consolidation of friable stone masonry has rarely been permitted on medieval buildings in Britain, although it has been employed in various ways on ornamentations and sculptured monuments. In general, where stone disintegrates it is usual to employ a mason to replace or reface it with some similar stone, if possible from the original quarry.

The removal of algae and other growths is usually done by spraying with an aqueous algicide. To this is often added a small proportion of sodium methyl or ethyl siliconate to render the masonry water repellent after the growths have been removed. A recently evolved process which is proving highly satisfactory, is described as the 'HVJ Double Cleaning Process'. This consists of cleaning by means of a high velocity jet, using a relatively small amount of water, followed by spray application of Thaltox (quaternary tin) algicide to protect the cleaned surface by inhibiting growths of algae, lichens and mosses. Where bituminous, sulphurous and other chemical deposits are involved, preliminary soaking, or the application of softening materials, is sometimes necessary, otherwise only plain water is used in the cleaning operation. Experiments to this end are being carried out with quaternary ammonium compounds, particularly on Carrara marble which 'sugars' easily and is particularly sensitive to careless or ignorantly devised treatments.

Reconstituted stone, plastics and synthetic resins are frowned upon by archaeologists and specialists in the preservation of ancient monuments and medieval buildings. Nevertheless, considerable work is being done with these materials, and epoxide resins are well to the fore in providing a durable and, if properly prepared, inconspicuous medium for patching and repairs. If mixed with brushings from the stone to be repaired, the insert being carefully covered with the same stone dust until the resin has 'cured', a very strong patch can be formed which will defy detection, especially if the whole area of wall is subsequently sprayed with silicones. If this is not done and the wall becomes damp, the impervious patches will become easily visible.

One of the oldest materials used to bind disintegrating stone is waterglass or sodium silicate. Potassium silicate together with many different additives is much discussed and much patented.

In the hands of the experts, they may have some value, but used indiscriminately may well do more harm than good. Fluorosilicates and siloconesters also require knowledge and discrimination in their use, for, if applied to certain stones or in certain circumstances, they may cause serious damage. A paper on stone preservation in Germany by Josef Riederer of the Doerner Institut in Munich, which was presented to the 1970 Conference of the International Institute of Conservation, describes modern materials.

Techniques for preserving ancient buildings have tended to come from empirical experiments designed to find the panacea for specific ills like spalling bricks, disintegrating mortar, peeling mural decorations and flaking tiles, all of which are directly or indirectly due to the penetration of water. Stone is, of course, subject to other problems. However, thermal movement, the effects of ultra-violet light and other factors not dependent on the presence of moisture would seem of insufficient economic importance to justify the cost and time of experimentation.

Theories and cures for dampness

General causes

Though we in Britain may feel that our weather is about the worst in the world, our houses are on the whole far better built and far better clad against the weather than those in most other countries. In actual fact, it rarely rains continuously for long in this country and twenty-four hours of uninterrupted rain are unusual.

Rain which falls or is driven on to the walls of buildings either soaks in or runs down the surface. If the material of the wall is completely impervious to water, then most of the rain will run down the drainage system, but, if it absorbs the rain, a vast weight of water may be built up inside the brickwork or masonry in a few hours. However, unless the atmosphere is completely saturated or the rain is heavy and driving enough to form a continuous film of water over the surface, the process of evaporation begins. In most instances, when rain wets a wall, evaporation is taking place even while the raindrops are falling, and the rate of evaporation rapidly increases as the rain diminishes and stops. Evaporation controls water penetration and, as a result, many houses built of extremely soft and porous materials are quite dry inside, while houses of much harder and less porous bricks or stone are damp.

If a material does not have to be impervious to be effectively dry, then, by the same token, the presence of moisture in buildings cannot be said to represent dampness in them. Indeed moisture may be present in a very large proportion of the materials which, together, constitute a building. Masonry may permanently contain an appreciable amount of water without being regarded as damp. Wood, also, may contain water up to 20 per cent of its weight, and still be regarded as 'dry'. Even biological organisms such as the dry rot fungus requires a moisture content of over 20 per cent in the wood to encourage spore germination.

Dampness in a building might be defined as the presence of moisture detectable by sight, touch or possibly smell. Staining of walls, ceilings or floors, coldness in the affected areas, or even a chilliness in the atmosphere provoke a feeling that the house is 'damp'. So dampness in buildings does not refer to any scientific condition of moisture content, but is a loose description of a condition of moisture which can be seen, felt or smelt.

The visible evidence may not be actual wetness but fungal moulds, peeling wallpaper, bulging, blistering and falling plaster, or the appearance of efflorescent salts on the walls. The detection of dampness by touch can be far from reliable unless the affected area is actually moist. Generally, a feeling of coldness is regarded as evidence of dampness but, unless this is extremely obvious, most walls, even in centrally heated buildings, are surprisingly cold to the touch. Rubbing the flat hand over a wall might reveal dampness, as the presence of moisture produces a dragging effect. The odour of dampness usually indicates the formation of fungal moulds or the musty smell which develops from damp plaster. The only sure way to judge whether a wall is damp is to use an electric or electronic moisture-detecting instrument. Few of these can be accurately described as 'meters', although they are often sold as 'moisture-meters'. If they are genuine meters, they will accurately register the amount or percentage of moisture present. Actually, most of these instruments detect moisture and give a general indication of amount within certain narrow limits.

Until 1935, when Starling and I began our study of dampness, the only methods of controlling it which had occurred to experimenters were all based on sealing the pores. However, we early discovered that dampness in a building was represented by *the difference between the rate of absorption and the rate of evaporation.* Anything that reduced evaporation without stopping absorption was likely to increase, rather than reduce, dampness. This simple formula guided all our further work. We reasoned that the mere fact of a substance being porous does not make it a conducting or permeable substance. It is the effect of the material on the surface tension of the water that decides whether it will conduct the moisture or resist or even repel it. Although glass is impervious to water, if it is formed into a very tiny bore tube, it attracts and conducts water in opposition to the pull of gravity. Conversely, some materials which physically appear to offer the best conducting system may be completely water-resistant and, despite their porous nature, will not admit water unless it is applied under pressure. A simple and general statement might be that material which readily absorbs water is a *hydrophilic* substance while that which resists or repels water is *hydrophobic.*

The majority of materials used in conventional buildings are hydrophilic. Bricks, stones, mortar, concrete, plaster, wood, paper, wall boards and most clay and quarry tiles are hydrophilic and, to a very varying degree, will take up water. Many other materials, particularly metals, glass and some plastics, are impervious without being hydrophobic. To make a porous and permeable material hydrophobic, it is generally necessary to treat it in such a way that the surface tension of the water resists entry into the pores of the material, instead of being attracted into these pores. A very permeable brick, such as an insulation brick, can be made relatively hydrophobic by being dipped into a thin mineral oil and allowed to drain thoroughly. It will be found that the oil has now changed the properties of the pores and, although they are still open, the water will no longer enter them; the repellency now produced will cause a tiny air bubble to form over each pore and seal it off.

Unfortunately, the hydrophobic effect produced by oil breaks down after a short time, and so has little practical application as a water-repellent or water-proofing medium for use on building materials. In fact, water repellency can be produced more effectively by using physical means rather than oils or similar materials. If a material is coated with very fine and very short hairs and fibres, it can be extremely hydrophobic. This is the way in which most water-resistant vegetables and some animals are protected from water making contact with their bodies or blocking the pores of their skins. The fine hairs, in a mass, are unable to break through the surface tension of the drops of water. The surface tension, in effect, forms a bridge between one hair and those surrounding it under which is a cushion of air. Even fabrics such as velvet demonstrate the peculiar effect of water repellency. If a little water is spilled on to velvet or plush, it will tend to break up into drops and lie on top of the fabric for several minutes before it starts to spread and soak down into the material.

When, in 1935, the first efforts were made to reproduce this effect on bricks and other porous materials, the experimenters, Richardson & Starling, hit on the idea of forming artificial hairs on and in the bricks by crystallizing paraffin wax in the pores. Some of their experiments were carried out by dissolving paraffin wax in various organic solvents, including solvent naphtha and various petroleum distillates ranging from heavy fuel oil to petroleum spirit.

Paraffin wax in petroleum spirit produced a fair degree of water repellency, but the wax formed a greasy sheen on the treated surfaces, tended to seal the pores and was, of course, highly inflammable. So similar experiments were carried out using two non-inflammable solvents, carbon tetrachloride and trichloroethylene. Of the two, trichloroethylene was the better but it still left a sheen on the surface on cool masonry. Only on masonry warmed by the summer sun was there no visible appearance of the treatment left on the surface. Many samples of wax were requested from every petroleum company in this country, but a wax with a melting point of 49°C, obtained from a German firm, proved to be the most satisfactory.

This wax was not actually soluble in trichloroethylene but, when melted, could be suspended in the heavy solvent and easily dispersed by gentle agitation. It was at first used at a strength of 10 per cent weight/volume in the solvent but, owing to the separation of the wax into a flaky scum on the surface of the trichloroethylene, the strength was reduced. Oddly enough, as the strength was reduced, so was the efficiency increased, until it was found to be considerably more effective at 5 per cent than 10 per cent and even slightly more efficient at 1 per cent than at 5 per cent. All was revealed under a microscope. At 10 per cent the concentration was too great for the crystals of wax to form properly. At 5 per cent they were formed but were very close together. At 1 per cent they formed in such a way that they were completely bridged by the surface tension of water, and an air space occurred between the water and the substrate. When brushed or sprayed on porous bricks, the fluid penetrated almost immediately and almost as quickly the trichloroethylene began to evaporate. As the last of the solvent disappeared, an invisible deposit of micro crystals was formed, reproducing in effect minute hairs. All the crystals were pointing outwards in the direction of the evaporation and immediately made the most pervious brick as hydrophobic as a lupin leaf.

Demonstrations of this particular treatment were impressive because the globulation of water sprayed on the treated brickwork was truly sensational—the water actually bounced from the surface. This process was first used as a treatment to render wall surfaces water-repellent. In 1938, it was adapted to provide horizontal and vertical damp-proof courses by drilling masonry.

Other processes, for instance the 'Knapen' process, have relied on increasing the area of evaporation without increasing the absorption. This is achieved by drilling fairly large holes, sloping upwards from the external face of the wall, and inserting porous, fireclay tubes of triangular external section with the apex of the triangle upwards. The porous tube is set in a very porous mortar mix so that the moisture rising in the masonry will be transferred to the mortar, and thence to the Knapen tube inside which the moisture evaporates. The evaporation causes a reduction in temperature, so the vapour flows down the tube and is discharged. Evidently Knapen acted on the same assumption about the cause of dampness in buildings as did Richardson & Starling.

Soon after the end of World War II, extravagant claims were being made for the virtues of silicones. Apparently, no car, floor or furniture polish was really satisfactory unless it contained silicones. The manufacturers strongly recommended certain grades of silicones for rendering porous masonry water-repellent. My son, Barry Richardson, who was investigating the application of silicones to building construction materials, very soon found that the recommended silicones were decidedly unsatisfactory for the treatment of limestone. He therefore carried out experiments with water-soluble siliconates which proved far more effective in rendering limestone water-repellent. Unfortunately, solutions made with these materials are caustic, having a pH value of about eleven. They also tend to change the colour of some forms of limestone, particularly those containing iron.

His investigations and experiments suggested that the silicones soluble in organic solvents remained in a liquid form, and did not 'cure' when applied to certain forms of masonry. This he eventually accounted for by the presence of bicarbonate of lime. This acidic radical, present in limestone, green concrete and mortar, was, in contrast, neutralised by the alkali of the water-soluble siliconates, so the silicone content was able to cure and become effective. As there was some doubt about the ultimate fixation of the silicones applied in aqueous solution, the first outcome of this investigation was to pretreat with the aqueous solution of siliconates, and to follow this up in a week or so with a further treatment of organic solvent silicones.

This idea worked well and some firms and organisations are still using this dual treatment. However, in 1957 and 1958, Barry Richardson set about combining, in one material, the virtues of both types of silicone. The addition of an amine group to the existing silicone molecule made it as effective on limestone, green concrete and mortar as on clay bricks, tiles and other non-calcareous substances. Thus the material most used as a water repellent for the long term maintenance of building materials was evolved.

Specific causes of dampness

Many buildings are erected with inherent faults which cause or promote dampness. Others develop faults as time goes on and general deterioration, movement or even biological influences produce conditions which result in dampness.

41

MASONRY AND DAMPNESS

Inherent faults:

It is impossible to enumerate all the built-in faults or omissions which result in the occurrence of dampness in buildings, but the following will help in the valuation of property and judgement of whether it is likely to be warm and dry or cold and damp.

1 Solid porous masonry; particularly that which is rendered, stuccoed, or rough-cast on the outside.

2 Absence of horizontal damp-proof courses inserted below the ground floor and above the soil level.

3 Absence of vertical damp-proof courses between the main walls and contacting brickwork or masonry.

4 Absence of impervious cappings and flashings to parapets, pediments, set-offs, string courses, chimney stacks, pinnacles and other ornamentations and excrescences.

5 Porous retaining walls.

6 Inadequate or non-existent provision for ventilating subfloor spaces, basements, kitchens and other places where atmospheric humidity can build up to produce condensation, also spaces between the ceilings and roof coverings of flat roofs, metal-sheathed cupolas and turrets etc.

7 Porous roofing materials, including some types of asphalt, concrete and stone shingles set at too low a pitch.

8 Porous concrete or stone lintels, sills, ashlars, quoins, etc.

Developed faults

1 Solid masonry which was originally waterproof but subsequently, through movement, has developed cracks and other faults, or fractures or deterioration of damp-proof courses.

2 Stone, and sometimes brick, which has become porous or absorbent as the result of the action of atmospheric acids, biological organisms (lichens, mosses etc), frost or the absorption of sea-spray or other soluble and hygroscopic salts.

3 Cavity walls with the wall-ties bridged by mortar, dropped by careless bricklayers or due to the falling of disintegrating mortar.

4 Cavity walls in which mortar slovens have accumulated at the bottom of the wall to a level above that of the damp-proof course.

5   Deterioration or the breaking away of flashings due to weathering and wind.

6   Weather, wind and frost damage to chimney cappings, cement weathering, fillets, tiles and slates.

7   Deterioration of lead and zinc gutters, valleys, etc due to mechanical wear, oxidation or the acid released by mosses and other vegetable growths.

8   The rusting and cracking of iron goods, separation of gutter joints, stopends, etc due to deterioration of screws and fixings.

Casual and created faults

1   The blocking of gutters, hopper heads and down pipes by leaves, birds' nests, balls, etc.

2   The heaping of coke, coal, firewood or soil against the walls.

3   Raising the level of the garden border or path above that of the damp-proof course or above the internal floor level.

4   The application of an impervious finish to the external surface of porous masonry such as water-proofed rendering, oil-bound or emulsion paints or a combination of cement rendering and decorative finish. Unless ventilation of the interior of the building is increased, such external finishes invariably increase internal condensation.

5   The erection of walls, brought up to and in contact with the main walls of the building without regard to the necessity of damp-proof courses to stop bridging or transfer of absorbed moisture from the new wall to the original building wall.

6   The planting of shrubs and trees close to walls, the effect of which is to reduce evaporation of directly absorbed rain-water or rising moisture.

Also the growth of trees adjacent to the house discharging water from their leaves on to the walls and causing movement by the expansion of their roots.

7   The use of salt on floors to keep down dust.

8   The placing of rain-water butts or tanks so that any overflow is discharged onto or against the wall.

The remedies

Mechanical damp-proofing

This list, although by no means comprehensive, may assist those faced with penetrating and/or rising damp to decide on which measure is most suitable for their particular problem. In most instances, the course of action is fairly obvious, but a true understanding of the cause is the first essential to success. Very often, modifications to the structure, the introduction or improvement of ventilation, or trenching and soil drainage will achieve all that is required.

Generally, modifications to the structure would mean taking down defective and leaking parapets and rebuilding them, with the addition of damp-proof courses, flashings and other built-in safeguards against penetrating dampness. Indeed, it might simply mean introducing a damp-proof course where none existed before.

In such a case, a conventional damp-proof course of metal, plastic or bituminous felt might be inserted into a continuous slot cut, horizontally, right through the wall, below floor level and above soil level. Although mechanical cutting machines are available, a specially designed handsaw can be used. Whatever method or equipment is chosen for the purpose, the cost is high and considerable skill is required to produce successful results.

It is not unusual for such a damp-proof course to aggravate instead of alleviate the problem of damp. If the solid brickwork of a house in an extremely exposed position has been dealt with in this manner, when the dampness was in fact due to rain-water penetrating the whole face of the wall, then this penetration will still continue, and moisture entering the brickwork will percolate downwards and build up on top of the damp-proof course. During prolonged periods of rain, particularly with driving wind, a calamitous state of affairs may develop. Water has been seen actually flowing out of the joints of such walls, along the top of the damp-proof course. Weep holes must then be cut into external joints to persuade the water to flow outwards. To forestall any such eventualities, however remote, it is wise to apply appropriate water-repellent treatment to the external face of the walls, particularly on the prevailing weather side.

Rising damp can also occur through faults developed in an existing course, but this usually results in patches of damp rather than a long continuous strip running along the walls. In many instances, where rising damp has occurred on walls with an existing course, it has been caused by moisture, taken into the wall above the level of the course, percolating downwards, and not by moisture rising in the wall and passing through defects in the course.

A simple method, successfully used to stop damp from rising in the walls of churches, is to dig a trench along the foot of the outside wall, making certain its depth is below that of the floor inside the church. The trench is filled with hardcore topped with shingle. If the trench is provided with ample soakaways, and rain-water discharge is conducted away from the walls so that there is no danger of the trench filling with water, the result is often highly satisfactory. It is essential, of course, to prevent the growth of weeds and to ensure that the hardcore and shingle does not become clogged with mud or other deposits. While the hardcore and shingle remain clean, the evaporation of moisture is increased to the point where it is able to cope with the absorption and so, to all intents and purposes, the interior of the wall becomes dry.

Chemical damp-proofing

Since 1938, walls have been successfully provided with damp-proof courses by being drilled and injected with solutions of wax, latex and resins. The first recorded method was to drill holes about $\frac{5}{8}$in in diameter, sloping *downwards* into the wall for about two thirds of its thickness. The holes were spaced at 9in centres and a measured quantity of paraffin wax solution was injected into the holes, until an amount equal to 1 gal of solution per 15cu ft of masonry was diffused. Walls of 13½in thickness or over were drilled from both sides so that the holes on each face of the wall alternated.

Unfortunately, the wax would not diffuse into masonry that was too damp or cold, thus creating many problems of application. Eventually, the treatment was rejected as being impracticable but was revived when water-diffusing siliconates became available. Apart from the development of a self-measuring irrigating bottle in place of the injection nozzle, the method employed by many 'damp-proofing specialists' is today exactly the same as in 1938.

One firm has patented a modification of this simple method by having rings of foam rubber around the perforated tube which is inserted into the holes in the masonry. This overcomes, to a great extent, the wastage which often occurs when the damp-proofing fluid runs into cavities and open joints, for the sponge rubber only transfers the liquid to masonry with which it is in contact. Unfortunately, this method means drilling large and often unsightly holes, which are difficult to hide without coating the whole area of wall with a rendering or distemper. Rendering a wall after the insertion of any damp-proofing course will often lead to a breakdown in the system due to 'bridging'. The rendering acts as a conductor, conveying moisture from the soil over the damp-proof course and into the masonry above. This modification, therefore, can create its own problems.

The use of water-soluble siliconates which diffuse into the moisture contained in the masonry does not deal with the problem of salts deposited in the wall by water rising from the soil. Such salts will often crystallize and form efflorescence, sometimes hygroscopic, on the interior face of the wall. To help overcome this problem, it is usual to remove all interior plaster which has become stained by rising damp. This is because most dissolved salts accumulate near the surface of the wall by the evaporation of moisture, thus causing the high concentration in the plaster which is generally the reason for the staining.

A more recent and more positive method of providing a chemical damp-proof course is by the pressure injection of silicone resins dissolved in petroleum solvents. The equipment consists of a compression pump to which one or more flexible pipes are attached (but not usually more than six). On each pipe is a metal injecting tube of length varying with the depth of penetration required. Each tube is fitted with an expanding gland which will seal it into the horizontal hole drilled into the masonry and thus enable the liquid to be injected at high pressure. Flow-meters are often incorporated to show the amount of fluid injected.

Since the injection pipe is narrow-gauged, only a small hole in the brick, stone, or concrete is needed. This can easily be filled and hidden with suitably coloured mortar or cement. One advantage of this method is the displacement of water in the wall by the organic solvent solution. The holes are drilled into the bricks, enabling each brick to be visibly saturated as the water, followed by the organic solvent fluid, oozes from the joints. This water is often visibly discharged from the masonry within minutes of the operation.

Cavity walls or walls filled with loose rubble are treated, either by drilling from both sides, or by first treating the external skin then, using a longer bit in the same holes, drilling right through each brick and into the brick on the other side of the cavity to a calculated depth. This can be marked on the drill, and should be about 3in into the brick itself or to two thirds of the thickness of stone walling. A longer nozzle is then inserted through the external skin and cavity into the inner skin, where it is sealed by the expanding gland as before. This has rapidly become the most popular process and is used throughout Britain and on the Continent.

In recent years electro-osmosis has been introduced to check rising damp though it has not been tested by the writer. One such process is based on the electrostatic charge which builds up in a damp wall and which can apparently be discharged by running an electrode round the building. This is set into the masonry and discharges the electrical potential into the ground by copper conductors. When the static has been discharged, the moisture level falls and the wall becomes dry. The other system of (*activated*) electro-osmosis depends upon the passage of an electric current from inside the wall to the outer face. Electrodes are set into the inner and outer faces and between them is passed a current at a low voltage. This is said to convey moisture from the inner to the outer face of the wall where it will evaporate. However, there is great danger of alkali corrosion of masonry at the cathode and acid corrosion at the anode, though this depends very much on the salts dissolved in the moisture contained in the masonry. Neither electro-osmosis process has been proved over a long period on medieval stone buildings in Britain, but sensational reports of the activated process have been received from Germany and the Balkans.

Whatever type of damp-proofing is chosen to overcome existing rising damp, it is essential to get rid of any hygroscopic salts which have been taken up with the moisture. These salts usually migrate and crystallize in the plaster on the internal face of the wall. When the atmosphere is humid, the crystals liquefy and give the appearance of damp, even if rising damp has been successfully checked. Removal of plaster which is visibly stained will, in the great majority of cases, remove most of the hygroscopic salts.

Walls of bare brickwork, masonry and concrete on which hygroscopic salts have crystallised, can be treated by extracting the salts with cellulose sponges. Large motor sponges, two buckets and a plentiful supply of warm water are the prerequisites of success. One bucket is filled with warm water. A very wet sponge is held to the wall for several minutes. During this time, some of the soluble salts will be transferred to the sponge by diffusion. The sponge is then squeezed out into the empty bucket. Five bucketfuls of water or about ten gallons are generally sufficient to extract the salts from a square yard of walling. Any form of injected damp-proof course should be applied after the salts have been extracted and the wall has dried out. This operation is obviously most successful during dry weather, or after a trench has been dug along the wall to reduce the level of the rising damp.

Once again it must be emphasized that to ensure success, the whole of the wall from footings to eaves should be sprayed with a suitable water repellent after the insertion of *any* type of damp-proof course. This is best applied to masonry that is reasonably clean and visibly dry. Before treatment, it is advisable to brush down the walls to remove dust and loose debris and to remove lichens and algae by the use of suitable thallogicides. (It is advisable to make sure the water repellent conforms with British Standard BS 3826: 1969.)

Cracks or faults in the masonry should be remedied and, where mortar or cement has been used, it should be allowed to dry. The silicone solution is then applied by means of a pneumatic horticultural spray at a pressure not exceeding 25 lb psi. Spraying should commence at the highest point of the wall and the spray nozzle should be moved horizontally across the face of the wall until the masonry at that point is saturated and liquid runs down the wall for several inches.

Windy or rainy days should be avoided and all glass and fresh paintwork should be protected. It sometimes helps to have a fugitive dye in the solution to show up the areas that have been treated. Once walls have been dressed with an effective water-repellent solution, they will not accept water-based paint or cement washes for about a year. This does not mean that the water repellent breaks down in twelve months. What, in fact, happens is that ultra-violet rays and other weathering factors change the molecular structure of silicones, but only to a depth of about $\frac{1}{32}$ in in the masonry. Although this has little effect on the efficacy of the water repellent, it enables water-based paints and cement washes to adhere to the surface.

Ventilation

The tendency to condensation has been greatly increased in recent years by the disuse of coal fires and the consequent sealing off of chimneys to stop draughts. This greatly reduces ventilation, and the problem is aggravated if the coal fires have been replaced by inadequately flued gas fires or oil heaters. Both of these forms of heating release a vast amount of water vapour which adds to that produced by other domestic machines. Even central heating introduced into an old, damp house can be very dangerous for a few weeks after installation, if the house is not kept well ventilated during this period. The heating quickly extracts the accumulated moisture from every porous article in the house and the atmosphere rapidly becomes saturated. Unless this moisture is quickly released into the outside air, it will become over-saturated and may precipitate or condense to produce exceptionally damp conditions in the cool roof space or other cool parts of the house. Some areas are certain to become warmer more quickly than others and in the thick-walled houses built over a hundred years ago, months may elapse before the whole building reaches a uniform temperature. When central heating is first introduced, all windows in every room should be opened just an inch or two at the top and bottom so that thermal currents are created and the hot, moisture-saturated air is expelled.

Most outbreaks of wet and dry rot are produced by excessive dampness due to condensation. Houses that have been free of timber decay for many years may, soon after a change of ownership, be attacked by dry rot generally because the new occupier had different ideas about warmth and ventilation from those of his predecessor. If the previous owner was fond of fresh air and always kept windows open in cellar, cloakroom and bedrooms, and the new owner 'hated draughts' and had all windows and doors sealed, then the change is bound to cause excessive dampness.

Occupied houses which are ventilated and warm stand little chance of developing fungal decay. The mere action of walking across a suspended floor may cause it to act like a diaphragm pump and create an appreciable movement of air in the room, cellar, or floor space beneath. But should any house be left unoccupied and untended for a few months, particularly in the winter, the stagnant air and consequential condensation which occurs during fluctuations of temperature, will often produce fungal growth in the form of moulds on walls and ceilings and, very much worse, dry rot fungus in the structural timbers.

Many wooden floors in old buildings will remain in good condition while they are left uncovered or only partially covered. If, however, the occupier suddenly decides to cover the floor with lino, tiles or some other impervious covering (and this includes rubber or plastic-based carpets and underlays), the passage of air through the joints between the floorboards is stopped and condensation with consequent decay soon becomes apparent. Draughty houses may be unpleasant and possibly unhealthy from the occupier's point of view but, so far as the house structure in itself is concerned, every draught is a blessing.

Ventilation by means of air bricks is often incomplete and unsatisfactory. The introduction of air bricks is often haphazard and without regard to the possibility of stagnant pockets in the corners of sub-floor spaces. Air bricks should be so placed that air will circulate to every part of the space. These currents are produced by various conditions, the most common being the difference in air pressure, due to wind, on one side of a building as compared with another, or the temperature on the sunny side compared with the temperature on the shady side. Where there is no wind and no sunshine, movement of air can take place by thermal means if, for instance, the temperature in the sub-floor space is above or below that in the outside atmosphere. Where possible, to encourage such thermal movement of air, the air bricks should be inserted at two levels, one level being as low as possible, the other just below floor level.

If the atmosphere outside the building is cooler than that in the sub-floor space, air will pass in through the lower air bricks,

Plate 5 A screwdriver penetrating a beam in which dry rot has been hidden by recent painting

Plate 6 Serious damage of a heavy pendant in a hammer-beam roof in a building frequented by many visitors. Only one rotted dowel was holding it

Plate 7 Merulius development on furniture stored in a damp cellar

Plate 8 Typical dry-rot growth found under floor covering

warm up, rise and then be expelled through the upper air bricks. Conversely, if the air outside is warmer than the air inside, it will pass in through the upper air bricks, cool down, fall and pass out through the lower air bricks. This will, almost certainly, ensure a movement of air at all times. It cannot, of course, be done where there is insufficient depth of brickwork between the soil and the floor. Honeycombed sleeper, partition and party walls are, very obviously, essential to ensure a free movement of air from one side of a building to another. Where solid floors abut on to suspended floors and so create a barrier to the free movement of air, ducting should be introduced into the solid floor to ensure free movement of air through the sub-floor space of the suspended floor. In many instances, the ducting through the solid floor has to communicate with the outside through a swan-neck vent. Such a vent should not be covered with a hinged flap or it will only permit air to pass outwards and not inwards. This could greatly reduce air movement at a time when the pressure on the outside side of the house was higher than on the other side where the suspended floor is situated.

Air grids at the side of open fireplaces greatly assist in extracting air from sub-floor spaces by passing it up the chimney, thus promoting ventilation of the sub-floor air and reducing draughts in the living area. As open fires disappear and central heating takes their place, a healthy situation can be encouraged by inserting grids under or in the skirtings behind the radiators so that cool air from below the floor can rise and be warmed by the transfer of heat from the radiator.

Gas radiators should always be vented to the outside air but, if this is done via an existing chimney, once used for coal or wood fires, care should be taken to ensure that the vast amount of water produced by the combustion of the gas will not soak into the lining and produce stains on the chimney breast of the room above. In most instances where gas heating or gas-fired boilders are introduced into an old house, it is advisable to insert impervious liners into the flues.

Chapter IV

Fungi in Buildings

The mushroom, *Agaricus campestris,* is possibly the best known fungus. There are, of course, many thousands of other fungi ranging in size from some so minute as to be invisible to the naked eye to others many square feet in area. Most fungi occur out of doors. Some spring from the soil and others attach themselves to trees, fences and other hosts which provide their nutrition. If they are growing on living hosts, they are known as parasitic fungi and if on dead organisms, as saprophytic fungi. The very small fungi, such as one finds growing on jam, cheese and stale foodstuffs, generally known as moulds, may also be found growing on damp wallpaper, plaster, emulsion paint and other damp substances in a house. The majority of these so-called moulds are propagated by means of microscopic spores. These are developed in a tiny sac known as an ascus, and so the group are called Ascomycetes. The spores so formed are ascospores but, in some species of Ascomycetes, other imperfect and asexual secondary spores are developed and these varieties of Ascomycetes are described as fungi imperfecti.

The fungi imperfecti may also propagate without the use of spores, by fragmentation of the fine threads known as hyphae, which can only be simply described as the microscopic roots which grow out of germinating spores, and which become longer and longer and by branching and sub-branching form a weft of tissue called the mycelium, or spawn. When propagation takes place by means other than by germination of normal spores it is known as vegetative propagation.

So far as the deterioration of buildings is concerned, the most important group of fungi are the Basidiomycetes, or those fungi similar to the mushroom. These grow their spores on club-shaped structures known as basidia, which are themselves borne on gills like those on the underside of the umbrella-shaped cap of the mushroom. The most common fungi found attacking timber in buildings come into this class, although they do not necessarily bear any visual resemblance to the common mushroom.

All the fungi which naturally occur in and on building materials and fabrics are saprophytes, and are performing the function for which they were designed — destroying dead organic matter in order to provide themselves with the means to live. This consumption of organic matter requires the presence of water, and so dampness is always associated with fungal growth in buildings as it is in fields and forests. The fungus actually consumes its host by releasing from its hyphal cells a digesting fluid. This enzyme acts upon the carbohydrate substrate and converts it into soluble substances which may then be absorbed by the hyphae in a similar manner to roots taking up moisture and foodstuff from the soil. In utilizing the carbon, hydrogen and oxygen of which carbohydrates are composed, the fungus not only produces foodstuff for absorption, but produces more water from the hydrogen and oxygen.

The substance of most decayable building materials and fabrics is cellulose. Indeed, it is the main constituent of wood, paper wallboards and insulation boards and is readily attacked by a wide variety of fungi if it becomes sufficiently damp to encourage the germination of spores. Some of these fungi consume the pale-coloured cellulose without consuming the dark-coloured lignin, the other important constituent of wood. They have thus earned themselves the name of 'brown rots' since the decayed wood goes brown. Other fungi, which attack lignin as well as cellulose and thus remove the darker element from the wood, are known as 'white rots' and, generally, reduce the wood to a fibrous white substance. The brown rots cause the wood to shrink and crack both longitudinally and transversely, so forming cubes or oblong blocks which become dry and may be easily broken off and crumbled in the fingers.

It has lately become the practice in scientific papers to refer to dry rot, previously known as *Merulius lacrymans,* as *Serpula (Merulius) lacrymans.* In this book, however, I am using the appellation which is still familiar to the majority of people who would commonly refer to the species of fungus called *Merulius lacrymans* as dry rot, and most other forms of timber-destroying fungi as wet rot. (The desiccated, charred appearance of the wood after basidiomycete attack probably gave rise to the term dry rot, by which, at one time, all fungi attacking timber in buildings was known.) Neither term is appropriate and, until more precise descriptive names are adopted, both are inclined to cause confusion. Both require damp conditions to promote spore germination and to encourage their growth and expansion but, in the case of *Merulius lacrymans,* the fungus, once established, can spread to relatively dry wood by producing its moisture requirements from the wood itself. In contrast, the so-called wet rots will only grow on wood which is moistened by some other means, such as roof leaks, defective plumbing, damp penetration through walls, rising damp or condensation. In the case of the wet rots, the growth and expansion of the fungus comes to a halt when the supply of moisture is cut off. *Merulius lacrymans* may continue to grow for quite a long time, providing the ambient atmosphere is still and can be kept in a high state of relative humidity by the water produced by the fungus.

If a house is damp, the temperature at which it is kept will have an effect on any fungus which starts to develop. In general, fungi grow more rapidly as the warmth is increased up to about 30°C. Then, as the temperature rises, the rate of growth falls off and, at about 38°C, it ceases. The fungus is not necessarily destroyed by this temperature but may survive for some time, particularly if ample moisture is present. If, however, the high temperature is accompanied by the removal of the source of moisture and the atmosphere and substrate become dry, the fungus will quickly die. Hence the warning, which cannot be repeated too often, that the introduction of central heating into a damp house may create a very dangerous situation unless accompanied by ample ventilation to ensure the quick removal of water vapour from the atmosphere.

Wet Rots

Coniophora cerebella

This is the most common fungus found attacking building timber in Britain. It is also found in the open and is responsible for economic losses to stacked timber, agricultural implements, building equipment and other wooden objects stored out of doors in a damp situation. It generally occurs in buildings where wood is in close contact with damp masonry, soil, oversite concrete, or where condensation occurs as the result of dampness combined with lack of ventilation.

Coniophora attacks hardwood more slowly than softwood, and many years can elapse between the initiation of the attack and the collapse of an affected piece of hardwood timber. Although known as the cellar fungus, this brown rot is very frequently found attacking timber in roofs, particularly the timber enclosed between the lead, and the ceiling attached to the underside of the rafters of a flat or shallow-pitched roof. Vapour, passing through the porous plaster of the ceiling and building up to a state of high humidity in the cavity, condenses when the lead on the roof reaches a lower temperature than the moisture-laden air. The boards on which the lead is laid quickly become saturated with water. This situation invariably results in the germination of *Coniophora* spores and decay of the boards. The dampness and the decay extend as the condensation continues and, in a surprisingly short time, softwood timbers can reach a state of collapse.

Even the lead may become affected by the combination of these various factors, for the moisture condensed takes up carbon dioxide, often released in large quantities by the heating used in medieval buildings. The moist condition and deterioration of the wood produce organic acids and these acting synergistically with the carbonic acid will act on the lead to form lead carbonate.

Coniophora often attacks building timber without revealing its presence on the surface. In general, this type of attack is due to internal condensation and results in the degradation and cubing of the inside of boards and beams, leaving an external skin of apparently sound wood. This may also happen to one side of the wood only. Condensation may have produced severe *Coniophora* decay on the underside of floorboards, while the top surface appears to be in excellent condition.

When signs of the fungus appear on the surface, they may occur as a thin yellowish-brown skin or as a dark-brown (sometimes almost black), group of strands which take up a vein-like pattern. The thread-like strands are, in fact, the mycelium or vegetative 'body' of the fungal plant and the sporophores or fruiting bodies are formed on the yellow skin. These take the form of rounded tubercles or irregular lumps, starting off as a creamy-yellow skin but turning to a dark, greenish brown as the sporophores are formed. Oak, chestnut and elm may show very little change of colour as the result of the attack, but softwoods become distinctly brown and, eventually, almost black. Concentrating on the cellulose and associated polysaccharides in the wood, the fungus secretes enzymes from its hyphae. The enzymes liquefy these carbohydrates, and hydrolize them into soluble glucose which the fungus can then absorb. This consumption of the substance of the wood causes it to shrink and crack, both longitudinally and transversely, thus breaking what remains into cubes which, when dry, can easily be pulverized between the fingers.

Coniophora can withstand wide extremes of temperature and has been known to survive temperatures ranging from $-30°C$ to $+65°C$. However, it does need a relatively high moisture content in wood to initiate its attack and extend its growth. It thrives when the moisture content is between 34 and 46 per cent of the oven-dry weight of the wood and the temperature around 23°C, conditions common enough in badly ventilated cellars and roof spaces, inside cupboards, under staircases and behind panelling, and found even in ancient buildings with very primitive heating arrangements.

The fungus is resistant or tolerant to many chemicals which, when used against other fungi, are effectively toxic. Zinc chloride, copper sulphate, with or without chromate, cadmium sulphate and sodium fluoride, all have to be used in relatively high concentrations to be effective and so are no longer used for remedial treatment. This is now carried out with chemicals such as tri-n-butyltin oxide, pentachlorophenol and orthophenyl phenol, dissolved in organic solvents like light aliphatic or aromatic oils. *Coniophora* responds best to typical wet-rot treatment. Firstly, any source of dampness from water leaks or rising damp must be eliminated by repairs. But *Coniophora* very often occurs as the result of condensation which can only be controlled by adequate ventilation. If, by these or other means, the affected wood is brought to a moisture content below 20 per cent of its oven-dry weight, then the fungus will die. If the wood has still sufficient strength to serve its purpose, it may safely be left without fear of any extension or revival of growth provided, of course, that dry conditions are maintained. As an additional safeguard, all replacement timber is usually pre-treated with a suitable preservative, and timber which, although affected, is sound enough to do its job, is treated 'in situ' with the types of fungicide already described.

Other wet-rot fungi found in buildings may be dealt with in the same way as *Coniophora cerebella*. Although a number of forest fungi may be found attacking the external and internal timbers of old buildings, these invariably occur as the result of dampness which has itself arisen from neglect and lack of maintenance.

Phellinus megaloporus (cryptarum)

This fungus is not uncommon in ancient buildings, as it is associated with the decay of the large oak and chestnut timbers found in such buildings. It generally occurs in roof spaces, church steeples, under lead gutters, in situations where rain-water has been able to penetrate and accumulate in deep cracks, shakes and joints and where the temperature can become unusually warm, as it does during the summer under lead roofs. Wood attacked by *Phellinus* becomes particularly attractive to the Death-watch Beetle and, very often, the appearance on the surface of holes made by the beetle is the only indication of the fungus working away in the centre of a great beam or plate.

The fungus may attack a beam for a number of years before the sporophore is seen. It is then often found filling and overflowing from the crack into which the water penetrated. Once seen, it is often mistaken for the *Merulius lacrymans* sporophore. There is, however, a very distinct difference. Whilst the sporophore of *Merulius* is damp, clammy and soft and, even when dried, has a soft leathery feel, that of *Phellinus* is hard and woody and buff or brown in colour. Indeed, it is often quite difficult to break it away from the wood. The attacked wood is reduced to soft, white, fibrous strands with the consistency of lint. When an affected beam is sawn through, it is often possible to pull out masses of this soft, fibrous and, often, apparently dry material which bears no resemblance to the oak or chestnut from which it was produced.

Just occasionally, *Phellinus* is found in a very active state, when it may produce a yellowish-brown matted growth on the surface of the wood from which drops of moisture exude. This gives it an even stronger resemblance to *Merulius lacrymans*, although the colour of the liquid from *Phellinus* is yellowish-brown whilst the 'tears' of *Merulius* are crystal clear.

Poria vaporaria (vaillantii)

This is another wet rot which is occasionally found attacking timber in buildings. It usually occurs where water drips consistently on one spot for a prolonged period, such as a leak from a parapet or valley gutter dripping on to a wall-plate. The mycelium appears on the surface of the affected wood as opaque, white sheets and branching strands. It will begin growing on wet softwood and spread to contiguous wet hardwood. It will not, however, extend beyond the actual area of wetness to attack drier wood. The amount of damage caused by this fungus is, therefore, relatively small.

Dry rot

Merulius lacrymans

Although not so common as *Coniophora cerebella*, dry rot has probably caused far more serious and extensive damage to buildings. Indeed, if the dry rot fungus had been named 'malignant cancer of wood', owners of affected properties, who often feel that unnecessary destruction takes place when dry rot is being dealt with, might more readily appreciate the violence of the infection and absolute necessity to cut out every trace of it to produce a cure. Like *Phellinus megaloporus (cryptarum)* the dry rot fungus, *Merulius lacrymans*, does not grow in the forest, but seems to have adapted itself to attack timber in human habitations.

The cause of dry rot in buildings is dampness, more often than not as the result of neglect of reasonable maintenance. In old buildings, inherent, long-established dampness in the walls, due to lack of damp-proof courses, moisture passing into basements through retaining walls, streams running under the buildings and so on, causes only moderate damage from the effects of wet-rot fungi. But, should a water vent from the gutters get blocked and rain-water run down the walls, making them unusually damp for a comparatively short time (but generally exceeding eight weeks), then *Merulius* spores are likely to germinate on any inside timber which is in contact with, or in very close proximity to, the damp wall. Again wood is very susceptible in cupboards, behind panelling, cornices, wainscots or other cavities where the air is still and the ambient atmosphere becomes saturated with water vapour.

Very often, the *Merulius* growth appears inhibited while the environment is very wet but, if this is remedied and the wall permitted to dry without opening up the unventilated timber, there comes a time when the hyphae of the fungus begin to extend and spread through the affected wood and into the contacting masonry. From there, it often attacks timber on the floors above and below, and extends through partition and party walls to attack the timber on the other side. Timbers built into masonry, such as bressummers, lintels, stud framing, bonding timbers, etc, are particularly susceptible and assist in transmitting the fungal activity from room to room or, in the case of terraced houses, from house to house.

Occupied houses in which conditions ideal for the development of dry rot have existed for years may not actually show signs of the fungus until, for some reason, the house becomes unoccupied and closed up. Such stagnant conditions often encourage a saturated atmosphere and, in a short time — perhaps only a few weeks — the mycelium of *Merulius* spreads and establishes itself.

The first signs of dry rot may take the form of small whitish, rubbery-looking growths along the edges of wainscoting, skirting boards, joints in panelling or from between floorboards, window linings, shutter cases. These are the incipient sporophores which expand and open up to reveal the hymenium that in its turn will produce vast numbers of reddish brown, air-borne spores. It may be the appearance of this reddish brown spore dust settling on objects in a house, blowing up through joints in the flooring or covering the shelves in cupboards, which first indicates the presence of this dreaded fungus.

As these sporophores only occur after the fungus has been consuming wood for at least several months, their appearance invariably indicates that appreciable damage has already taken place. To reveal this damage necessitates the removal of anything which may be hiding it and the growth which caused it. The consequent stripping of joinery, plaster, tapestries, pilasters and other magnificent and beautiful decorations found in ancient buildings is both a highly skilled and, at the same time, highly depressing task. Often, the superficial indications of dry rot are quite small and, to the uninitiated, quite insignificant but further investigation will reveal growth extending for many feet in all directions.

To attack and consume the wood, the hyphae which first penetrate, release enzymes which break down wood cellulose leaving lignin which becomes changed by the chemical action to a brown, easily crushable substance. The consumption of a major proportion of the wood causes shrinkage, and the brittle nature of the remaining lignin produces cracks which divide the decayed residue into cubes easily separated and crushed between the fingers. As the hyphae extend they branch and weave together to form the mycelium.

In dark, dank cellars and sub-floor spaces, the mycelium can exist in white flocculent masses like loose cotton wool on which can be seen drops of water (hence *lacrymans*), produced by the fungus. This water is the result of metabolism and the conversion of hydrogen and oxygen which, together with carbon, are the chemical constituents of wood or other carbohydrate materials. The carbon is combined with oxygen to form carbon dioxide. Thus, wherever *Merulius lacrymans* is active, the substrate on which it is growing and the ambient atmosphere are kept moist with H_2O, and acid with CO_2.

Conducting strands extend from the mycelium which can penetrate masonry and the soil under floors. From these strands, specialized hyphae produce the enzymes and acidic conditions which assist in this penetration. Upon reaching another piece of wood, these conducting strands or rhizomorphs produce further masses of hyphae which cohere to form more mycelium.

When the active mycelium is revealed, patches of yellow and mauve coloration quickly appear, and, as the humidity of the atmosphere surrounding the fungus is reduced, the snowy-white mass shrivels and clings more closely to the wood or substrate over and on which it is growing. Eventually, it takes the form of a mouse-grey coloured skin, still tinged with small areas of yellow and mauve. This skin may be peeled off and remains soft and flexible even when dry, but the rhizomorphs or conducting strands become quite brittle when dry and may be snapped with the fingers.

It is only after the fungus has been left undisturbed for a year or so that the sporophores begin to develop. These usually show up on the surface of an affected area within the building, but occasionally may appear on the face of the wall outside, particularly where shrubs or other growth have protected it from sunlight and wind. Like many fungal growths, *Merulius* sporophores are most likely to appear towards the end of the summer and during the autumn. In an advanced state of growth, they have been described as 'pancake-like' in their appearance, the edges of the growth being greyish white and the centre reddish brown. The reddish brown area is the hymenial surface from which millions of spores are released. Everything in its vicinity will become coated with brick-red spore dust.

The sporophores may vary in size from an inch or so to several square feet. One sporophore discovered on an arch of a closed and redundant church measured over ten feet in length. Those which form on a flat surface remain flat like a pancake but, on a vertical surface, they may become bracket-shaped with the hymenial surface developing a stalactite appearance. The odour of a freshly growing outbreak is very similar to that of mushrooms but, as it becomes old, decomposition takes place and a most unpleasant odour is produced.

No matter how small the sporophore, the surveyor wishing to discover the extent of the infection and damage should open up an area of at least 100sq ft for his preliminary examination and then extend his investigations until he reaches the last trace of mycelium, from which point he will continue for another 3ft. This opening up of the building fabric must extend if possible, in all directions. Examination should include, depending on where the outbreak occurs, the ceiling or roof above and the floor and sub-floor below, the walls (both sides), and rooms to right and left. The surveyor must not be even daunted by the prospect of dismantling a valuable staircase, fireplace, mural decoration, carving or anything behind which the fungus may possibly have spread. Those who have not been so ruthlessly thorough, have invariably regretted their squeamishness. The final cost of dealing with unchecked dry rot has escalated to double, treble or even more than the original outlay, had courage not deserted them in the first place.

Removal of the cause of dampness, complete exposure of every inch of infection and growth, excision and removal of all affected wood and very thorough treatment with suitable fungicides are the essential requirements for eradicating an outbreak of dry rot. Even so, the liberated spores will be present in every part of the building and, given suitable conditions, may germinate and start a fresh outbreak. It is therefore essential to examine the whole building and investigate any possible cause of dampness. Wherever possible, ventilation should be introduced into every stagnant cellar, roof space, room, cupboard, staircase spandrel, cavity behind panelling, cornice, coving, soffit or any other place where the combination of damp and stagnation might give rise to a further outbreak of *Merulius lacrymans*.

Chapter V

Insects in Buildings

For centuries the choice of roofing material played an important part in the length of time that buildings remained habitable or useful. In most parts of Britain, reed, straw, heather or wooden shingles were the most favoured materials. They were readily available, inexpensive, warm, waterproof and, above all, required only a very light timber frame to support them. Where thatch was used, this light framework usually consisted of sapling trees, trimmed to form poles, often with the bark still adhering. Unfortunately, the use of saplings and young hardwood trees led to attack by the Common Furniture Beetle, which has a preference for sapwood. As this form of framework generally contained more than 75 per cent of sapwood, the damage caused by these tiny insects was, in most cases, calamitous. No matter how well the thatch itself was maintained, the insects would still attack the supporting timbers so that the roofing structure often collapsed.

Danger from fire, particularly in towns, gradually led to the use of more fireproof types of cladding such as stone slabs, tiles or slates. In some parts of the north of England, large and heavy sandstone flags were used, in the Midlands and the South, smaller limestone tiles and in Wales and Cornwall, local slate. The significant feature about the replacement of thatch and shingles by heavier forms of roofing material was the effect it had on the choice of supporting timbers. Beams and rafters of much larger dimensions had to be employed. The proportion of sapwood was appreciably reduced, and so were the effects of *Anobium* infestation.

However, the use of more substantial timbers depended to a large extent on money, and the disappearance of many small buildings must be attributed to the damage caused by the

Common Furniture Beetle, generally known as 'woodworm'. It attacked not only roof timbers but also the wattle in the walls, the elm and beech flooring and any other part of the building constructed of or dependent upon hardwood or softwood containing a high proportion of sapwood. In those days, when the timber most used for construction was oak or chestnut and often came from dead trees or branches, the combined action of fungi and the Death-watch Beetle was mainly responsible for the decay and disintegration.

Modern houses may be subject to similar forms of attack, particularly by the Common Furniture Beetle and a fairly wide range of fungi. The Death-watch Beetle is unlikely to be the culprit in modern buildings unless second-hand hardwood timbers from old buildings have been used in their constuction. Many pseudo-Tudor houses, for instance, have incorporated old beams, staircases or panelling to make their appearance more convincing.

Fungi, both dry rot and wet rots, may cause disastrous damage in a short time where dampness is allowed to persist, but insects will attack the timber even when it is not damp. Of the woodborers, it will already be apparent to the reader that the most prevalent is the Common Furniture Beetle or 'woodworm', and that the Death-watch Beetle confines its activity, almost entirely, to oak and other hardwoods in ancient buildings. The next in importance, from the viewpoint of a property owner or custodian, is the House Longhorn Beetle, *Hylotropes bajulus*, the Powder Post or Lyctus Beetle, the wood weevils and several beetles with no common names such as *Ptilinus pectinicornis* and *Ernobium mollis* as well as, for those living near water and dockyards, the Wharf Borer, *Nacerdes melanura*.

The Death-watch Beetle (Xestobium rufovillosum)

The Death-watch Beetle probably earned its common name during the Middle Ages. Its characteristic ticking noise was heard, generally in the still of the night, by those sitting with the sick and dying. It is surprising how far the sound produced by this tiny beetle, which is less than a centimetre long, can carry in a still house, being audible at twenty feet or so if the tapping occurs on some hollow or resonant container.

This tapping, made by both sexes, is the prelude to mating

which only takes place when the atmosphere is warm. Ovipositing does not take place for several days, during which the female carefully explores the surface of the wood, seeking a suitably fungal-infested or decayed part. The eggs are usually laid in small groups of three to twelve though, exceptionally, groups of eighty or more may be found. They are translucent white, oval and have a somewhat pointed lemon-shaped end. About ⅓mm wide and ½mm long, they can easily be seen with the naked eye. The incubation period varies considerably and is greatly influenced by temperature. Thus, during a warm spring (the eggs are laid from the end of April until June), the eggs could hatch in three weeks, but in cold weather the period may be extended to five weeks. Under a hot lead roof, in conditions of high relative humidity, the period may be as little as two weeks.

The freshly emerged larvae can only penetrate decayed wood and often begin by passing down a vessel. As the wood is consumed, so the gallery or tunnel which they create becomes larger until, just before pupation, it may reach a diameter of ³⁄₁₆in.

The larva is characteristic of the anobiids with an enlarged thoracic region, a narrow 'waist' and a further, but slightly less enlarged region made up of the last three abdominal segments. The body is curved and, when it gnaws its way in the wood, the rear dorsal end of the larva is pressed against the wall of the gallery to give purchase, so that the mandibles may be forced into the wood. The larva is thus practically bent double inside the wood.

Pupation occurs after a varying period of three to ten years. In the late summer, the larva excavates a pupal chamber about ¼in from the surface, closes the gallery with fragments of wood and forms the creamy coloured pupal cuticle in which metamorphosis to the imago takes place. This period of metamorphosis varies from three to four weeks when the cuticle is shed. The beetle then remains inside the pupal chamber in a comatose state until the following spring and, when the temperature rises, emerges by gnawing a perfectly circular hole. The wood removed by the mandibles is not ingested, but is pushed back underneath the body. When the beetle emerges, these wood fibres, together with the bore dust or 'frass' often fall from the flight hole. This bore dust or frass may be easily identified since it occurs as perfectly circular, bun-shaped pellets, so hard that they feel like coarse sand. Vast quantities of this frass may be released when a cavity excavated by the larvae is opened up.

The insect is reddish brown with patches of short, stiff, yellow hairs on the elytra and prothorax. When the beetle is freshly emerged from the wood, it often has wood dust adhering to the hairs, which gives it very effective camouflage. It remains dusty for several days and often until it dies, so the reddish brown colour is usually hidden. A common beetle in the forests of Europe and in England, it is most frequently found in the decaying stumps of the pollarded willows so common on the banks of rivers and streams. Unlike other beetles of the anobiid family, the Death-watch Beetle is not a vigorous flier.

Although the larvae may bore in the wood for a period of three to ten years, the adult beetle rarely survives for longer than six weeks. The length of time spent by the larva between emerging from the egg and becoming a pupa seems to be inversely proportional to the extent of the decay caused by fungi. Consistent dampness with steady fungal activity reduces the life cycle of the Death-watch Beetle to a minimum. Where the decay is caused by Basidiomycetes, maturity is reached faster than in those cases where Ascomycetes and fungi imperfecti have caused the damage. In fact it is in those instances where the wood has been attacked by fungi, stimulated spasmodically by condensation alternating with drying-out periods, that the life cycle is prolonged. From my own experience, the minimum life cycle, in conditions like those under the floors of most churches in southern England, was three years. This was in oak attacked by *Coniophora cerebella*. In the roof timbers of Winchester and other cathedrals, the most concentrated areas of Death-watch Beetle attack were discovered where the fungus *Phellinus megaloporus* (*cryptarum*) had developed. As these circumstances and conditions reduced the length of the life cycle, so the speed at which the insects multiplied increased, and heavy concentrations of flight holes drew the attention of the custodians to the infestation.

Where fungus and beetle attack is confined to sapwood, the number of flight holes visible on the surface may be more alarming than significant. In general, such areas of attack may

Plate 9 A Common Furniture Beetle on a cocktail stick

Plate 10 Damage caused by the Common Furniture Beetle

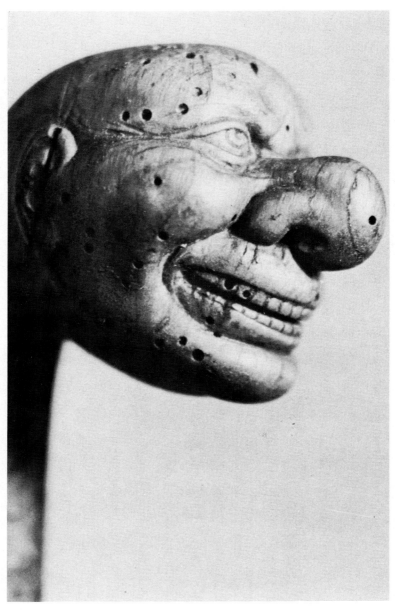

Plate 11 Head of a walking stick damaged by the Common Furniture Beetle

be trimmed away, leaving the heartwood in good condition and causing no appreciable weakness to the structure. But, very often, the moisture which has given rise to the fungal attack finds its way into shakes, faults, splayed knots and joints. This enables the fungus to spread its hyphae into the heartwood, leaving the outside of the beam or post apparently sound. As the beetles follow the penetration of the fungal growth, the centre of such timbers is gradually excavated and many, on inspection, prove to be shells or hollowed-out boxes, sometimes only a fraction of an inch thick and filled with frass. Such a beam or post does not slowly buckle as its strength decreases but suddenly collapses, releasing gallons of gritty boredust. The collapse of one might well overload other weakened timbers in the vicinity, resulting in a general collapse.

When the attack is confined to sapwood, the full extent of the damage can be seen and its seriousness assessed. However, when the beetles succeed in forming a cavity deep inside the wood, they can emerge into that cavity, lay eggs and begin fresh life cycles without showing the extent of their activities on the surface of the wood. It is a very unwise surveyor who expresses an opinion on the state of a timber structure where any evidence of Death-watch Beetle activity has been discovered. He can just as easily overestimate as underestimate the damage and to assert, without actual proof, that the beetles are active can be foolhardy. There have been many instances where timber in old buildings has been found riddled with the flight holes of the beetle, but the most careful and competent investigation has produced no trace of a live insect. This situation seems to occur where a building has been neglected for a long time and then central heating and good ventilation have been introduced. In such circumstances, it seems that either the beetles are discouraged or that the new conditions are more favourable to their predators than to the beetles themselves. The most common of these predators is another beetle, slightly smaller than the Death-watch and bright blue in colour, with the name *Korynetes caeruleus.*

Common Furniture Beetle (*Anobium punctatum*)

Anobium punctatum is the commonest of the anobiids and, although of European origin, is now distributed throughout the temperate regions of the world. Although appreciably smaller than the Death-watch Beetle, being only $1/10$-$1/16$ in long (2.5–5.0mm) a written description makes it apparent that *Anobium* bears a resemblance, morphologically, to its close relative *Xestobium rufovillosum* — the Death-watch. The prothorax forms a hood which covers the head, so that when one looks down on the insect the antennae protrude from beneath the prothorax and appear to be attached to it. The colour is uniform over the body and 'varies from light reddish-yellow through dark chocolate-brown to pitchy red'. This range of colour is thought to be due to the rubbing off of the yellowish pubescence, revealing the reddish-brown wing cases (elytra). These are covered with small pits in rows running longitudinally and are easily seen with a moderately powerful magnifying glass.

Like the Death-watch Beetle, it lays its eggs in small batches of two, three or four, in cracks, often on the end grain, but always where they are likely to be safe. The joint between one piece of wood and another is a favourite spot. The wood does not have to be decayed by fungus as in the case of the Death-watch Beetle, although wood containing less than 12 per cent moisture is rarely selected for ovipositing. In general, the beetles are attracted to areas of moderate dampness but not wetness, where the air is still. Extensive damage is often discovered in cupboards, under floors, in cellars and behind panelling, where none was expected.

In ancient buildings, the sapwood of oak is generally used for decorative carvings. This is frequently found to be riddled by the borings of *Anobium* and so fragile that the slightest tap or movement will cause it to collapse. Panelling, cornices, balusters, newel posts, bench ends, pulpits and hundreds of other decorated objects, often beautifully carved, have perished and disappeared as the result of *Anobium* activity.

Oak was the timber most favoured for supporting and surfacing floors, but elm and beech were also used extensively, especially when restrictions were placed on oak, which was used from Tudor times until the middle of the nineteenth century for building naval ships. Beech and elm are woods greatly favoured by the Common Furniture Beetle, and the former is often uniformly attacked with no differentiation between sapwood and heartwood. In many old houses, the main rooms were floored in oak, but the less important rooms and the servants' quarters were floored with elm, beech or any other timber which happened to be on the estate. This timber was not necessarily felled and sometimes came from trees which had died or been blown down in storms or struck by lightning. If the wood lay on the soil for even a few months before being used, it would almost certainly become infected with fungal spores or infested with insects before it entered the building.

In most instances, the presence of the Furniture Beetle is revealed by the appearance of small heaps of light-coloured bore dust. This powder is much finer than the bore dust of the Death-watch Beetle but feels 'sandy' when rubbed between the fingers. When examined by means of a powerful lens or low-powered microscope, the pellets are seen to be elongated or cigar-shaped.

The flight holes from which the insects emerge vary in size from 1-2mm in diameter. It is not always easy to decide if the holes are recently formed or have been present for years. As in the case of the Death-watch Beetle, the sight of holes in wood is no proof of current activity. In general, if fresh bore dust is present, it is reasonably safe to assume that the beetles are active. Proof, however, can only be obtained by the sight of beetles emerging, or the discovery of larvae or pupae within the wood. Likewise, a piece of wood, with no signs of a flight hole or other indication of infection, may have eggs deposited in cracks or crevices where even the most meticulous examination would fail to reveal them. Larvae which have not yet reached the emergent phase may be actively gnawing away inside the wood.

If the Furniture Beetle is definitely found to be active in one or two areas of timber in, say, a roof, then all similar timbers in the structure must be regarded as suspect. In any case, the fact that part of the roof structure is affected, suggests that the remainder of the timber has reached a condition where it has become receptive to the insects, though this condition varies with different woods. In softwoods, the attack may commence within ten years of the erection of a building but, in hardwoods, particularly oak, the period is generally much longer. This receptiveness may be dependent upon chemical changes in the wood or upon the development of symbiotic organisms, like micro fungi, which develop as the result of conditions of dampness. Certainly, the greatest incidence of Furniture Beetle damage in ancient buildings has occurred where the moisture content of the affected timber must, at times, have become high as the result of the intermittent and localised heating which invariably causes condensation during cold weather. Stabling, cow-byres and buildings in which animals are kept suffer timber damage by *Anobium* to an even greater extent than dwelling houses and churches. This can be attributed to the release of moisture from the animals, together with the formation of nitrogeneous materials such as urea and ammonia which increases the nutritional value of the wood.

There is a great diversity of opinion on the environmental and nutritional needs of *Anobium*, and no incontrovertible statement has yet been made to finalise these issues. The incontestable fact remains that the Common Furniture Beetle was, and still is, one of the serious causes of deterioration in buildings, not only to the structural timbers but, as the name implies, to the furniture and timber decorations in such buildings.

Wicker chairs, baskets and other furniture are also particularly susceptible to Furniture Beetle attack. They have often been responsible for spreading the infection, particularly when placed in a loft, store room or even a summer house, where they are often stored along with other highly susceptible articles. The beetle is encouraged to breed in these odd corners and can then attack the rafters and structural timbers.

The answer to this source of trouble is *never* to accumulate any wooden articles in a place where they are unlikely to be used or looked at for months or years. If they are valuable, then protective treatment should be applied before they are put away. If they are neither valuable nor useful to their owner then they should be given to someone who can use them, for articles in use stand far less chance of being infested with insects than those which remain stationary and untouched for long periods.

House Longhorn Beetle (Hylotrupes bajulus)

Great publicity was given to this insect when, soon after World War II, the Health Inspector at Camberley in Surrey reported finding evidence of heavy and active infestation in numerous council houses. Although instances of House Longhorn activity have been found in a number of other places in Britain, they have been isolated and, except for the north-west Surrey area already mentioned, there have been no further discoveries of concentrated infestation.

The adult beetle and the larvae are quite different in appearance and size from the other woodborers commonly found attacking building timbers in this country. In fact, only the relatively rare Oak Longhorn, *Phymatodes testaceus*, and the Wharf Borer, *Nacerdes melanura*, might be mistaken for this beetle. Even this is very unlikely, as the circumstances and situation in which they are likely to be found would serve to distinguish them.

The adult female of the House Longhorn species is usually very much larger than the male, the largest female being up to 25mm (about 1in) long while the male may be only 7mm long. The antennae appear thin and sharp pointed and are roughly half as long as the body of the insect which is 'greyish-brown to black and covered with greyish pubescence' (Hickin). Looking down on the insect, the most striking features are peculiar markings which, on the pronotum behind the head, look like two black eyes, a V-shaped nose and a mouth and, on the wing cases, look like two white commas. If infested wood is cut or broken and the larvae exposed, they will be found to have straight bodies which taper slightly from head to 'tail' and are segmented into twelve distinct sections. The holes from which the insects emerge from the wood (flight holes) are oval and vary from 3 - 9mm in length.

As surprisingly few flight holes occur in relation to the amount of damage caused by the larvae of this insect, the presence of a single hole will certainly indicate that a considerable length of galleries exists below the surface of the wood. It is not safe, therefore, to dismiss any sign of infestation as only a trivial occurrence. It is much safer to assume that any sign of current or recent activity is an indication of danger, to be thoroughly probed and investigated.

Powder Post or Lyctus Beetle (Lyctus brunneus)

New hardwood flooring, panelling, furniture or even picture frames may quickly show signs of attack by what appears to be the Common Furniture Beetle. In the spring, tiny heaps of dust and holes appear, all apparently the work of this most common of domestic woodborers. If, however, an adult beetle is found emerging from the wood and examined under a powerful lens, it may be found to be somewhat different from the Common Furniture Beetle. It will appear longer in relation to its width and, looking down on it from above, a distinct head, thorax and wing-cased body will be apparent, in contrast with the *Anobium* beetle which shows only the thorax attached to the body, the head being hidden beneath the thorax.

If the heap of bore dust feels soft, like talcum powder, and does not have the sandy or granular feel of Common Furniture Beetle frass, the damage is probably being caused by a species of *Lyctus* beetle. There are several other characteristics which will help to identify them and these, together with the above, may be summarised as follows:

They confine their attacks to hardwoods only. These hardwoods may come from any part of the world, but the species of *Lyctus* which attack wood grown in the temperate regions are usually different from those which attack tropical woods. Not all hardwoods are susceptible, for only those with large pores, such as oak, ash, elm, walnut, African mahogany and obeche, provide the insect with sufficiently large vessels in which to insert their eggs.

Only the sapwood is attacked. Starch is an important part of the *Lyctus* diet and occurs mostly in the sapwood of recently felled trees. *Lyctus* beetles do not attack standing trees or freshly felled logs.

Only partly or recently seasoned timber is attacked. As the wood seasons, the starch content is reduced until insufficient remains to sustain *Lyctus* infestation.

Heavily attacked wood tends to disintegrate to reveal the packed bore dust with little or no clear-cut indications of galleries, as in the case of *Anobium*.

The emergence holes are circular and rarely have a diameter of more than 2mm. These may occur in varnished or veneered surfaces and there are instances of the *Lyctus* beetle boring holes through lead and silver.

Stacked hardwood in timber yards is most vulnerable to

attack. In many recorded instances, the emergence of *Lyctus* from flooring, panelling, furniture and sports equipment, occurring very soon after installation, has been traced back to an initial infection which occurred in the timber yard or storage premises in which the timber has been kept before conversion. *Indications of Lyctus damage may often be found on ancient timbers.* But this is past history and no current activity will ever be seen.

Owing to the serious incidence of *Lyctus* damage which occurred in furniture soon after the end of World War II, control measures were introduced. Compensation had to be paid to purchasers of infected furniture and considerable publicity in the press led to much more careful storage, handling and kilning of hardwoods. The result was that *Lyctus* infestation has now become relatively rare.

Pinhole Borers or Ambrosia Beetles (*Bostrychidae*)

The damage caused by these beetles can be mistaken for that of the Common Furniture Beetle or the *Lyctus* beetle although, in fact, it has nothing like the same significance, neither does it cause collapse or disintegration in the manner of *Lyctus*. The size of the holes is, as the name implies, very tiny — about the diameter of a pin — but the holes are made more conspicuous by a dark staining of the rim due to the gallery being lined with fungal mould on which the larvae feed. In fact, pinhole borers are not wood-consuming insects in the manner of the anobiids and lyctids, for the tunnels into the wood are made by the adult female as a safe depository for her eggs. This occurs when the wood is green and, often, very soon after felling, particularly in the tropical forests where their activity may cause serious problems.

Timber damaged in this way is often used only for the inner layers of plywood which will be hidden by the final external layer. The strength of the wood is not significantly reduced by the presence of holes made by the Pinhole Borers but, of course, it restricts the use of such wood to those situations where a perfect appearance is of no great importance. The depreciation in value of logs infested or damaged by pinhole-boring beetles can be of economic importance. In recent years, spraying newly felled trees in tropical forests with insecticide has become an increasingly common practice. Back in Britain, the presence of holes in timber delivered to the timber yard or the final customer need cause no great alarm, and the wood need be rejected only where its appearance is particularly important.

Ptilinus pectinicornis

This insect has no common name. In appearance, it resembles both the Common Furniture Beetle and the *Lyctus* beetle, but is easily identified by the characteristic antennae. Those of the male are pectinate or comb-like, while those of the female are serrate. In general, the imagos are larger than those of *Anobium punctatum*, but, as both species vary in length, this is of little value as a guide. I have only on rare occasions found evidence of damage caused by *Ptilinus* in ancient buildings and, in every instance, the insect was found attacking beech which had been introduced relatively recently. In one case, it was noticed in a beech floor which was about forty years old, in another in beech joists which were not more than fifty years old and had been used for repair works under a fifteenth-century oak floor. In a third case, it was found in panelling which was about a hundred years old.

No significant damage by *Ptilinus* has been reported in this country, but there have been serious cases in Germany and central European countries.

Ernobius mollis

This insect might easily be mistaken for the Death-watch Beetle, being a member of the same family, the *Anobiidae*. It is, however, slightly smaller than the Death-watch Beetle and slightly larger than the Common Furniture Beetle. Unlike either, it is generally found attacking the cambium or growth layer of wood, immediately below the bark of softwoods. It is, therefore, only normally discovered attacking fresh building timber to which the bark is still attached. In some cases the larvae may penetrate deeper, but rarely more than ½in below the bark. It is, therefore, often described as the 'bark borer', but, as there are many insects which bore into bark to lay their eggs in a protected environment, the use of this name could be very misleading.

I have included *Ernobius* because it has, on many occasions, caused serious misgivings in the minds of architects and owners, who have discovered evidence of activity in softwood timber in new houses. It has usually been discovered by the discharge of bore dust, the emergence of beetles and the sight of bore holes or galleries very similar in size and appearance to those of the Death-watch Beetle. If the observer is sufficiently knowledgeable to notice that the timber is a softwood, is quite fresh and has the bark attached, he will realise that the insect responsible could not be one of the common destroyers of building timber but something unusual and, in fact, comparatively harmless.

Indeed, the greatest danger it is likely to produce is that of mistaken identity, which might lead to unnecessary and expensive treatment with insecticides. Trimming off all bark which may still be adhering to the wood, usually along the arrises, is all that is necessary. This can be done with a draw-knife, small axe or adze. However, attached bark indicates the presence of a high proportion of sapwood which, as previously explained, is very susceptible to attack by the Common Furniture Beetle, and must be regarded as very inferior timber.

Phymatodes testaceus

This beetle is sometimes referred to as the Oak-bark Borer or Oak Longhorn, and has caused considerable and quite unnecessary alarm to architects, builders, timber merchants and even to the surveyors of ancient buildings. The latter, when examining the oak or chestnut timbers of old buildings, often find that men constructing them did not remove every bit of bark and, subsequent to installing the timber in the building, the *Phymatodes* beetle had laid its eggs beneath this bark. The emergence of the adult insects produced holes which are quite conspicuous and have the characteristic oval shape of a Longhorn Beetle. Once there, these holes remain long after all activity has ceased. *Phymatodes* beetles are members of the *Cerambycidae* or Longhorn family and lay their eggs only under the bark of recently cut trees and recently dead, standing trees. Any sign of activity in a building need cause no apprehension that there will be further infestation or any weakening of the timbers. As in the case of *Ernobius mollis,* it is only necessary to cut away the bark to effect a complete cure. The wood in this case will, of course, be oak, chestnut or some other hardwood, and rarely pine, spruce fir, larch, or any other softwood.

Wood-boring Weevils

Those readers, interested in gardening and familiar with the weevil family of insects, will know that they are characterised by what appears to be an elongated snout, on each side of which are attached the antennae.

The antennae of most beetles point forwards or curve backwards. Those of the weevils, however, grow from the snout backwards towards the body of the insect and then, about half way along their total length, are bent almost at right angles to point forwards. Apple blossom weevil, pea and bean weevil and wood-boring weevils all have this peculiar appearance. In the British Isles, we have two species of wood-boring weevil which commonly occur and may cause anxiety. Actually, they are really of secondary importance as they only infest wood already attacked by fungal rot, and usually in situations where dampness is the primary cause of the deterioration. Almost any wood subjected to prolonged dampness will, sooner or later, be attacked by fungi which then produces the conditions favoured by these insects.

In appearance, the two species are so similar that only an experienced entomologist is able to distinguish between them. The indigenous species is *Pentarthrum huttoni*. The other species, *Euophryum confine*, has been introduced during the last fifty years and is regarded as a native of New Zealand. They are very tiny, usually no more than 4mm (¹⁄₆ in), long and reddish to dark brown in colour. The galleries excavated by these insects generally run with the grain of the wood and often between the annual rings or, in plywood, between the veneers and laminates. The wood thus separates into very thin sheets which can be peeled off to reveal the insects in all stages of their life cycle. The frass is very fine and often hangs in clusters with a snuff-like appearance, particularly from the sides and undersides of floor joists exposed in a damp cellar of floor space.

These signs of weevil activity point to a damp atmosphere, arising from a build-up of moisture released by the site soil or concrete or from the walls, compounded by a lack of ventilation and causing a fungal growth to attack the wood. It is, therefore, a visible warning that a serious situation has developed, or is about to develop, due to a weakening of the flooring timber. If the joists are very substantial and are still strong enough for the job, then the drying out of the atmosphere will often check the deterioration. Treatment with a combined insecticide and fungicide will ensure quick eradication, and is essential in those situations where the ventilation cannot be improved.

The Wharf-borer (*Nacerdes melanura*)

The name Wharf-borer was given to this insect in North America, because it was associated with damaged piling, groynes and other timber in docks and jetties. In Britain, it has been found attacking damp and decaying timber in many other situations, sometimes well away from ports, rivers or canals. However, it is most commonly found attacking damp and decaying wood in buildings close to harbours, wharves, estuaries and other places with saline water. There are reports of its discovery around the base of lavatory pans and of its larvae at the base of telegraph poles and fences on which dogs had urinated. Wood wetted with salt water or urine does seem to be more susceptible.

Wharf-borers vary from 7 - 12mm in length and are softer than one expects a beetle to be. Hickin describes them as 'yellowish brown, but the apices of the elytra are black. The antennae of the male have twelve segments, whilst in the female there are eleven only. The eyes, sides of the thorax, legs and ventral parts generally are blackish and the whole body is covered with dense yellow pubescence.'

The Wharf-borers could never be mistaken for wood-boring weevils although their habits are very similar and infestations are limited to damp, or even wet, decaying wood. So they, like the weevils, are indicators that the wood they are attacking is already at an advanced stage of dampness and decay.

Other Insects Commonly Mistaken for Wood-borers

Many insects which come in from the garden are brought in on firewood; these consume our food, clothes, carpets, wallpaper and even us, and can be mistaken for wood-borers or, conversely, wood-borers may be mistaken for them.

However, 'Any black or dark brown beetle more than half an inch long can be dismissed as being very unlikely to be a wood-borer.' Such black beetles common in houses include the Ground Beetles which are harmless but sometimes invade houses in such numbers as to be a nuisance. The common household pests most likely to cause alarm are Spider Beetles, the Bread Beetle, and Plaster Beetles. Larder Beetles and the German Cockroach are often confused with wood-destroying insects. (For brief descriptions of these, see Appendix I.)

Chapter VI

Surveying for Timber Decay

The drawings at the beginning of this chapter illustrate various building timbers *in situ* and are intended to help readers having no specialised knowledge to follow the text.

I have included in this chapter some of the instructions which are used in training timber surveyors in my own company, which deals with every type of building from a castle to a cottage. Obviously few readers, unless they happen to be connected with the industry, will be likely to undertake the type of survey described on their own properties. It will, however, give them a clear idea of how and where to look for potential trouble, an important point, and of the relative significance of different findings. I also felt that it would be useful for home owners to understand just how much work is involved in carrying out a timber survey conscientiously. These days, far too many 'surveys' are no more than a quick glance at a few timbers, followed by an estimate for treatment. I hope the details which follow will help to convince readers that the cheapest estimate is not necessarily the best or the most accurate.

I would also like to emphasize that treatment for dry rot should *always* be entrusted to professionals. Straightforward 'woodworm' and some forms of dampness treatment can be handled quite satisfactorily by home handymen. Dry rot is a different proposition altogether. Once established, it has the facility to spread in any and all directions. It is absolutely essential that the whole of the infestation is assessed and that treatment is extended to provide a sterile band beyond the limits of the attack.

Until stripping of plaster, panelling, joinery etc has been completed, it is impossible, even for an expert, to gauge accurately the extent of the outbreak. A partial treatment may remove the immediate and visible signs but, unless complete

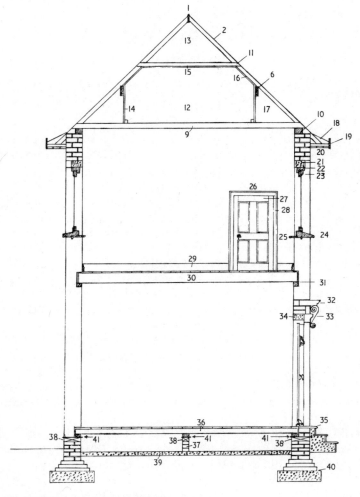

Fig 1 The structure of a house

Key to Figs 1, 2 and 3

1	Ridge	8	Sole plate	15	Ceiling (attic)
2	Common rafter	9	Tie-beam	16	Skilling (sloping) ceiling
3	Principal rafter	10	Wall plate	17	Spandrel space
4	King-post	11	Collar-beam	18	Fillet
5	Queen-post	12	Roof space	19	Facia
υ	Purlin	13	Cock loft	20	Soffit
7	Strut	14	Partition (attic)	21	Lintel

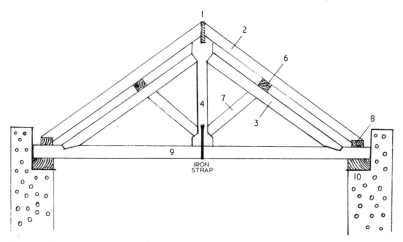

Fig 2 King post-truss

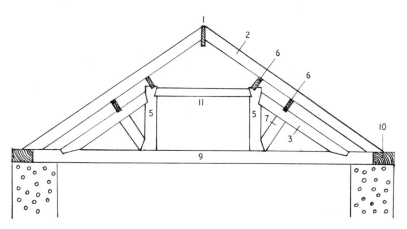

Fig 3 Queen post-truss

22	Window head	29	Skirting boards	36	Ground floor joist
23	Window rail	30	Joist	37	Sleeper wall
24	Sill	31	Plate	38	Damp-proofing
25	Window board	32	Moulded stone door head	39	Concrete oversite
26	Door head	33	Moulded corbel	40	Foundations
27	Top rail	34	Concrete lintel	41	Sleepers
28	Door architrave	35	Threshold		

eradication is effected, further outbreaks are almost inevitable. This may result in the destruction of new and expensive decorations before a successful treatment is finally applied.

Diagnosis

General requirements

Anyone who undertakes a survey of a building for the purpose of detecting and assessing the damage caused by insects or fungi will need:

1 Sufficient knowledge of entomology and mycology to be able to diagnose the form of decay, identify the species of organism, and know its characteristics and possible effect on the structure.

2 Some knowledge of other types of insects and fungi found in buildings which might be mistaken for timber-damaging organisms, but which, in fact, are harmless to wood or the structure of the house (see Appendix I).

3 Knowledge of building structures, fabrics, design, drainage, insulation, ventilation, heating and mechanics of construction.

4 Ability to make reasonably accurate sketches and plans for the purpose of illustrating reports, defects and possible structural repairs, modifications or reinforcements.

5 Ability to use a moisture meter.

Procedure for Roofs

1 Whatever the problem, type or size of building, always make a careful examination of the outside before going inside.

2 Always make an orientated, rough sketch plan of the roof structure before going into the roof space. This sometimes means going on to the roof, as it is not always possible to see the full extent of gables, valleys, etc from below. If the roof is large, divide the plan into sections, number each section and use corresponding numbers in the report. By this means, it is unlikely that any part of the roof will be missed or overlooked.

3 When going into a roof space, always note the level of the ceiling. Check whether there are any sloping ceilings or spaces behind which are hidden the wall-plates and lower ends of rafters. Never be satisfied until the wall-plates have been seen, particularly if under parapet or valley gutters. Always remember

Plate 12 Death-watch Beetles on infested timber

Plate 13 House Longhorn larva

Plate 14 Furniture Beetle damage to a lovely carving

that the most vulnerable parts of roof structures are those under gutters, and those buried, or partly buried, in masonry.

4   The timber carrying wide valley gutters or areas of flat roofing may be impossible to examine without removing the roof covering or ceilings. If so, say so in the report; never miss anything out of a report merely because it cannot be examined.

5   Always examine with a powerful torch or hand lamp which projects a beam of light. The ordinary lighting generally provided, or a lamp on an extension lead, does not disclose the small holes made by Furniture Beetles so clearly and easily as a concentrated spot of light, particularly when shone at an acute angle along the timber being examined. A sharp beam of light shone across the surface of timber, throws into relief the undulations caused by House Longhorn activity and the bright, clean edges of fresh flight holes of most wood-borers.

6   For probing the wood, use a small screwdriver sharpened to a fairly keen cutting edge. A round, pointed probe will produce holes which may be mistaken for insect flight holes by subsequent examiners. To test the larger dimensioned timbers for depth of penetration of insects or fungal attack, use an auger bit with a short length of steel piping as a twisting handle.

7   If clues on the outside, such as evidence of overflowing gutters, fall pipes, defective parapets and sills, built-up flower beds, etc suggest that dry rot may be present, then concentrate on its discovery before starting to explore for insect or other forms of damage. Always give priority to dry rot and make every effort to locate all outbreaks and the full extent of each.

8   If the building is less than ten years old, and 'woodworm' has been reported, anticipate finding *Lyctus* in hardwoods and *Ernobius mollis* in softwoods.

9   In roofs constructed entirely of softwood and built within ten years of the inspection, the most likely form of attack, as mentioned above, is *Ernobius mollis,* which can be effectively dealt with by cutting away all the timber to which bark is adhering to a depth of ¼in. The fact that this insect is present usually indicates that inferior timber with a high proportion of sapwood has been used. Such wood is very vulnerable to attack by the Common Furniture Beetle.

10   If a softwood modern roof has been damaged by fire and swamped with water, look for dry rot along the plates and

wherever timber is in contact with the masonry or is likely to trap water.

11   In softwood roofs over ten years old and particularly in those over twenty years old, the most common form of damage will be that caused by the Common Furniture Beetle, with *Poria* species of fungus where rainwater has leaked through faults in the roof covering or gutters. These forms of deterioration will be found throughout the British Isles.

12   The other serious pest of softwood timber roofs is the House Longhorn Beetle. This can do far more damage in a much shorter time than any other wood-borer. Although damage by the House Longhorn is mostly confined to a fairly clearly defined area of north-west Surrey, it is being found more and more frequently in other parts of the country, including Scotland. It is not easily detected until a considerable amount of damage has occurred, for the insects can work below the surface of the wood for as long as twelve years before emerging. The flight holes are often few and far between, and may occur on the hidden faces of rafters, joists and purlins, etc. The flight holes and piles of bore dust, which is usually very pale and soft, are much larger than those made by Furniture Beetles and the shape of the holes is usually oval. A skilled and experienced surveyor can detect the hidden workings of House Longhorn by lightly running his fingers over the surface of the wood. The tightly packed bore dust, usually only about $1/16$–$1/32$ in below the surface, causes a humping of the thin veneer on the surface and forms long, very slightly raised ridges. If a ridge is detected with no flight hole at the end of it, carefully dig along with the tip of a knife—it is quite usual to find a larva at the broadest end of the disclosed gallery, a definite proof of activity.

13   Flight holes, discovered in modern hardwood structures less than twenty years old, are generally those of the *Lyctus* beetle. This damage, confined to sapwood, is rarely of structural significance in roofs unless there is a preponderance of sapwood. However, it often spoils the appearance to an alarming extent, the significance of which can be grossly exaggerated, but, where such timber is visible, as in a church, college hall, library or similar building, the defacement caused by *Lyctus* can be disturbing. Its restriction to sapwood makes it obvious that, where serious damage has occurred, the quality of wood has been very poor.

14   If hardwood, particularly oak, has bark adhering, or has been cut very close to the bark, it is not unusual to find oval holes resembling those of the House Longhorn in the outer sapwood. These are not the work of the House Longhorn, which confines itself strictly to softwood, but of the hardwood longhorns, probably *Phymatodes testaceus*. These cause only superficial damage and leave the wood when the bark is removed.

15   Old hardwood roofs suffer many forms of damage particularly if, during their life, they have been neglected. Beech will often be found riddled through and through with Furniture Beetle or the closely related and very similar beetle *Ptilinus pectinicornis*. Oak, elm and chestnut often show signs of attack by almost all these insects with the exception of the House Longhorn. The Death-watch Beetle often seems to concentrate on the most vital and inaccessible parts of the roof structure. This is because it attacks wood which is visibly or invisibly infested with fungus, and the fungus grows as the result of damp due to direct wetting, contact with damp masonry, or by condensation. Usually therefore, the most concentrated areas of Death-watch Beetle attack are found under leaky gutters or cold gutters which condense the warm moist air in the building; along the ridge where moist air accumulates and condenses — at the ends of tie-beams and along wall-plates; in brackets, corbel posts, lintels and all timbers contacting or buried in solid or rubble-filled masonry or brickwork.

16   *Roofs of churches.* When examining the open roofs of churches, make a particular point of examining the purlin ends built into gable walls, rood arches, towers, etc, also the plates and gutter bearers, particularly those under the valley gutters over arcades and under parapet gutters. This often means the removal of boarding or plaster fixed to the ashlar posts. Look also, for plates buried in the top of walls, particularly the short ties and locking plates between the inner and outer wall-plates.

The wall posts, knees or bracket posts under hammer beams and tie beams, usually carried on corbels and either half buried in the masonry or flush up to the masonry, are also very prone to decay due to the dampness of the walls. The bases of king and queen posts, mortices and tenons often trap rain-water that leaks through defects in the roof. The water encourages pockets of concentrated fungal and insect attack, often only detected by drilling with an auger. Whenever staining has occurred as the result of damp, make a particularly close examination of the timber, for damp is always the forerunner of decay.

17 Where ceilings are fixed to the underside of the roofing timber, or are so close to the roof timbers that it is impossible to crawl between, the surveyor should insist on having the ridge and the parapet and valley gutters opened up to disclose the hidden timber. If they are found to be sound, it is reasonably unlikely that the timber between is defective.

18 When examining vaulted ceilings, particularly fan vaulting, always remember that it is the groins which collect all debris, insects or water which fall from the roof. Therefore, these are most likely to encourage beetle and fungal activity.

19 In churches with ceilings, never make the error of mistaking the ceiling for the roof. Wherever there are panelled ceilings or plaster ceilings, insist on seeing what is above.

Procedure for Interior Timbers

1 In the normal run of modern dwellings, the common forms of attack likely to be found below roof level and down to the basement are the Common Furniture Beetle, wet rot and dry rot, with the exception of north-west Surrey, where the House Longhorn Beetle is the most serious problem. Wood weevils are often found where fungal decay has occurred.

In the type of dwelling and in buildings where there are hardwood floors, beams, panelling and decorative timber, the *Lyctus* beetle and Death-watch Beetle may be added to this list.

2 Pinhole and shothole borers often cause unnecessary alarm but are not, in fact, serious destroyers of building timber. In most cases, the damage has been caused before the timber has been cut and prepared and, whether the timber is kiln or naturally seasoned, the attack has almost certainly ceased before the timber is inserted in the building. The holes of pinhole and shothole borers can generally be easily identified by the dark ring and staining round the edges and the absence of bore dust in the galleries.

3 Roofs of old stables and cattle byres are particularly prone to insect attack, usually by the Common Furniture Beetle. This seems to be due to dampness caused by the condensation of vapour given off by the hot animals.

4   If clues on the outside such as evidence of overflowing gutters, fall pipes, defective parapets and sills, built-up flower beds, etc suggest that dry rot may be present, then concentrate on its discovery before starting to explore for insect or other forms of damage. Always give priority to dry rot and make every effort to locate every outbreak and the full extent of each.

5   Having located an area of dry rot infestation, remember that it can spread in six directions; up, down, to right, to left, forward and backwards. Always examine adjoining rooms. Look in the ceiling and room above, the floor and room below, in and through the wall and up inside a stud-framed or brick partition wall. It rarely fails to penetrate into and through the walls, whatever they are made of, and fixing blocks and bonding timbers often provide it with nourishment and a means of spreading rapidly behind plaster and wall coverings over a wide area. If the growth occurs on a party wall separating the property you are examining from the adjoining property, ask permission of the adjoining property owners or occupier to examine the other side of the wall and the timbers in the vicinity in his property.

6   If a sporophore has formed, it is as well to assume that about 100sq ft of mycelium growth is hidden somewhere. Wherever it is, in the walls, under the floor, behind the plaster or panelling, it is up to the surveyor to discover it. If opening up facilities have been refused, record in your report that a sporophore was located and that a considerable area of hidden fungal growth is suspected. The only clue may be the characteristic red spore dust visible along skirtings, the floor joints between boards or in cupboards. This dust indicates that a sporophore is present, probably below the floor or behind wall-linings, and indicates an extensive growth somewhere in the vicinity.

7   In opening up, it is not unreasonable to extend the investigation at least 6ft beyond the last visible signs of growth, for the fungus will frequently disappear into a wall and reappear several feet away, the hyphae and rhizomorphs being hidden inside the masonry.

8   If it is necessary to lift floor coverings, cut away plaster, remove panelling or spoil the appearance of anything in order to make an inspection, find out before doing so who accepts the responsibility for making good.

9   Ground floors and basement floors, particularly if covered

with lino tiles, imitation parquet or other impervious material, are very prone to decay by the cellar fungus — *Coniophora cerebella*. Quite often this fungus can be present without appreciably changing the appearance of the wood. Only when the wood is probed, is it found to be seriously decayed, the signs of decay being hidden by a thin skin of apparently sound wood. Thin, brown or blackish coloured threads, rather like dirty cobwebs, stuck to the timber, may be the only indication of decay. Where cellar fungus is discovered, it will invariably be found in damp, inadequately ventilated spots and caused, not so much by direct contact with wet walls, soil, etc as by condensation of moisture and absorption of moisture from very damp air. This dampness is usually due to evaporation from a damp site or from porous walls, and the inability of the over-saturated air to escape. Quite frequently, sub-floor and cellar timbers attacked by the cellar fungus have little balls of snuff-coloured dust attached to them and peculiar runs of small holes along the grain of the wood. This is often mistaken for Furniture Beetle activity, but is in fact the work of wood weevils. From a surveying point of view, these insects are of little significance, as they only occur in wood already seriously damaged by fungal rot.

10   One common fungus found in old hardwood roofs is *Phellinus megaloporous (cryptarum)*. This is often discovered in oak which has been subjected to prolonged wetting by rain-water drips from a leaky roof or gutter. It generally germinates deep inside a shake crack or joint, and will grow through the centre of a beam for years without disclosing itself on the surface. A heavy concentration of Death-watch Beetle holes may be the only visible indication that fungal decay is present. When a sporophore appears and the growth spreads to the surface, it is often mistaken for dry rot. As it is rough and corky to handle, it can easily be distinguished from the soft and flabby feel of *Merulius lacrymans*.

11   When opening up floors over 150 years old to look for signs of decay, always try to locate the main beams into which the joists are notched. These main beams are the vital carcassing timbers and, being buried at each end into the walls, are extremely vulnerable. Always, therefore, remove a board against the wall over each end and test with an auger bit, even if the surface appears sound.

Chapter VII

Basic Methods of Timber Treatment

A very large number of different chemical formulations now exist to control fungal and insect activity in timber. With the exception of the 'mayonnaise' emulsion-type materials, most of these formulations are applied in such similar ways that standard methods of application have developed. These methods, given due consideration of all the circumstances and factors involved in each case, will provide effective treatment. Although these methods have changed very little over the last thirty years, there have been developments in ancillary work, such as the treatment of masonry and soil into which fungal growths have spread. But it must be emphasised that the thoroughness, conscientiousness and technical knowledge of those applying the treatment is quite as important as the method of application and efficacy of the chemicals employed.

On those occasions when disputes arise about the quality of the treatment, it is relatively simple to judge the materials by the formula. But, at present, it is harder to assess the efficiency of the method of application unless it can be compared with some official standard method which is known to produce satisfactory results. At present there are no such standards.

The apparatus for applying the preservative chemicals can be of the same type used for horticultural purposes. It should consist of a storage cylinder, in which the chemicals can be subjected to a pressure of three or four atmospheres by means of a pump. The pump can be operated either by hand, mechanically or by means of an inert gas from another high pressure gas cylinder. This pressure will force the chemical fluid through a control tap connected with a hose (made of material unaffected by the solvents), to which is attached a lance with a lever-control valve and fitted with a spray or injecting nozzle as required. A built-in and accurate pressure gauge is an essential part of this equipment. Air-blown, vaporizing, or paint sprays are unsuitable for the application of wood preservatives.

Where practicable and reasonable, depending upon aesthetic considerations and, particularly, upon the requirements of the owner or custodian of the timber to be treated, all extensively damaged wood should be trimmed away or removed. All timber judged to be too weak for its purpose should be brought to the notice of the owner or his representative, and recommendations made for its replacement.

Where, for any reason, the following described methods of application cannot be applied, a written explanation should be supplied to the client, preferably before the treatment is commenced. These methods are generally employed by the majority of firms and organisations engaged in remedial treatment throughout Britain, and, over a prolonged period, have given satisfactory results.

Treatment of Fungi

The application of chemicals for the purpose of controlling wood-decaying fungi in building is unlikely to achieve satisfactory results unless carried out by experienced operatives working to a specification, or under the direction of someone acquainted with the mycological, chemical and structural problems involved. Successful treatment is closely bound up with the knowledge and skill of the surveyor or inspector responsible for diagnosis. He must determine the type and species of fungus, the location of the limits of the infection and the specification of the remedial measures and control treatment to be adopted. The reasons why fungal spores have germinated and the decay has spread should be carefully checked and all faults which have led to or contributed to the outbreak should be remedied.

The following generic terms will be used in the text: *dry rot,* applied only to *Merulius lacrymans; wet rots,* applied to all other species of Basidiomycete and larger Ascomycete fungi which cause decay of timber in buildings; *moulds,* applied to superficial growths of tiny fungi such as *Aspergillus niger* and *Penicillium sp; soft rots,* applied to micro-fungi imperfecti.

Merulius lacrymans

As has been emphasised repeatedly, it is absolutely crucial to locate and reveal the full extent of the growth. The opening up and stripping of fabric should be extended to at least one metre beyond the area in which the growth can be seen. In situations where timber is buried or partly buried in masonry, concrete or soil, the growth may extend for many feet beyond any superficial indications of its presence. Such timber must be uncovered and investigated where there are any indications of fungal infection in the vicinity. This particularly applies to bressummers, lintels, plates and bonding timbers.

When all infected timber has been revealed and all other infected superficial building fabric removed, down to the basic wall or flooring material, the following procedure should be followed.

1 Remove all infected timber. Where a beam, joist, rafter or binder is only partly affected, the whole length of the timber need not be removed but it should be cut 18in beyond any visible signs of growth or decay. These directions do not apply to timbers which are built-in, or partly built-in, or which contact masonry, concrete or soil along their length. These timbers should be entirely removed, even if only a small proportion of the wood is visibly affected by the fungal growth. All wood thus removed should be carefully and scrupulously cleaned up, including sawdust, chips and decayed fixings, and moved to where it can be safely burned.

2 The affected masonry, concrete and soil should be brushed free of all fungal growth and the brushings removed to a spot where they can be safely destroyed. A blow-lamp or flame-thrower may be slowly passed over the affected area of wall or floor to consume superficial growths. However, this is not essential and cannot be relied upon to destroy fungus which has penetrated more than 3-4mm into the fabric.

3 Affected walls more than 4½in (12cm) thick should be drilled to produce holes at least ⅝in (16mm) in diameter, at 9in intervals up and across, sloping downwards into the wall fabric at an angle of 45° and penetrating to within 2in of the reverse face. Walls over 9in (23cm) thick should be drilled from both sides. External walls of any thickness need only be drilled on the

inside face unless there are visible signs of fungal growth on the external face. All dust from the drilled masonry or concrete should be carefully removed, preferably by vacuum machine.

4 Every hole should then be filled with an accepted water-soluble fungicide through a suitable nozzle on a spray machine, or by funnelling or the use of special equipment. The application should be repeated until a predetermined volume of fungicide has been absorbed. The fungicide should be equivalent in fungicidal toxicity to a 5 per cent solution of sodium pentachlorophenate and should be applied at the rate of at least 10 litres per cubic metre (35 cubic feet) of masonry. (All chemicals are now sold in litres but timbers and many measurements in the building trade are still in feet.)

5 The wall should then be generously sprayed with approved water-soluble fungicide over the whole affected area and 18in (46cm) beyond, in every direction, where that is possible.

6 Soil beneath ground floors and basements, over and into which dry rot mycelium and hyphae have passed, should be removed to a depth of 6in (15cm), over an area extending 3ft (88cm), beyond the visible infection. It should then be taken to a spot where it can be safely released or sterilised by means of heat. The area from which the soil has been removed should be generously sprayed with approved water-soluble fungicide at the minimum rate of 10 litres per cu metre (35 cu ft) of remaining soil.

7 When the fungal growth is on a partition or party wall, then the other side of the wall should be treated with the same thoroughness as the visibly affected side. If for any reason this is impossible, a full explanation should appear in the report.

8 All the acoustic and thermal insulation material laid on insert boards between floor joists (pugging), which are affected by fungal growth, or suspected of being affected, should be removed to expose the full depth of the joists. All affected pugging should be destroyed away from the building. Appropriate treatment, as described, should then be applied to the affected timber and surrounding fabric.

9 Finally, all timber of any description within 6ft of the visibly affected area should be thoroughly and completely treated with an organic, solvent type of fungicidal preservative approved by the British Wood Preserving Association.

10 All replacement and repair timber should be thoroughly treated with a persistent and approved timber preservative (either by vacuum pressure, using water-borne preservative salts, or by double-vacuum, dipping or steeping according to standard practice approved by the British Wood Preservative Association). Where the wood is cut or drilled, or the surface broken in any way to reveal untreated wood, all such wood should be thoroughly dressed by brushing or spraying with a suitable and approved preservative.

Wet rots

The most common of the fungi found attacking building timber and known generally as 'wet rots' are:

1 *Coniophora cerebella,* cellar fungus.
2 *Poria vaillantii* and other species of 'pore fungus'.
3 *Lentinus lepideus,* the stag's horn fungus.
4 *Paxillus panuoides.*
5 *Phellinus megaloporus* (or *cryptarum*), the oak fungus.

They all occur under damp conditions but none has the ability to penetrate into and through masonry in the manner of *Merulius lacrymans.* Consequently none of the wet-rot fungi extends its activity beyond the area of the dampness which promoted its germination and development.

As a result, the treatment for these fungi is much simpler and apart from timber or cellulosic material, requires far less opening up and stripping of contacting and adjoining plaster and building materials.

As in the treatment applied to control dry rot, the causes of germination and development of the fungus should be checked and remedied. Most outbreaks of cellar fungus are due to condensation of moisture in sub-floor or roof spaces, which are inadequately ventilated. This may mean structural modifications to provide additional ventilation and these, where necessary, should be recommended or carried out.

All affected timber should be tested to ascertain the degree of penetration of the fungus and to assess its remaining strength. Timber decayed to the point where it can no longer safely fulfil its function should be removed and replaced with adequately treated timber. Other timber which is affected, but is still sufficiently strong, should be treated by careful and comprehensive spray application of an organic, solvent type of fungicidal preservative approved by the British Wood Preserving Association.

Where insects as well as fungi are attacking the timber, an approved combined fungicide and insecticide should be used.

In general, the control of wet-rot fungi is ultimately achieved by removing the cause of dampness, followed by adequate drying out. Treatment with fungicide is necessary only as an additional safeguard and to remove the danger of a revival, or the development of dry rot, should the timber again become damp. Timber once affected by fungal decay is generally more susceptible to re-infestation.

In recent years the rotting of external joinery timber, particularly window frames, has caused considerable concern. The fungi which attack the wood usually develop from spores carried into cracks and open joints by the penetration of water, in the form of rain on the outside, and condensation on the inside. The decay often commences within two to five years of the erection of the building and is so prevalent at the present time that many thousands of windows are being replaced.

There are a variety of reasons why this is now occurring, although it was regarded as a comparatively rare problem in the past. The causes range from the use of timber containing a high proportion of sapwood, inadequately dried and carelessly stored before installation; poor design and fabrication; failure of adhesives or the breakdown of paint and other protective finishes. Joints are often cracked by slamming or forcing windows to open or close, and glass is often not effectively sealed into the frame. Even a tiny hair crack will permit moisture to enter and accumulate beneath paint or varnish and, where it reaches the exposed end grain, it will penetrate deep into the wood. Fungal spores germinate in the persistently damp conditions and cause softening and the eventual collapse of the wood.

Remedial treatment of such decay has recently become possible by the invention of a method for injecting fungicidal preservative into the joinery *in situ,* without removing paint or varnish, and at a low cost compared with replacement. This invention consists of a small, flanged, plastic tube with a connecting nipple at one end which is a non-return valve (see Plate 17). The treatment process is applied by drilling a hole near the joint, with a suitable bit, to a predetermined depth in the wood and inserting the plastic tube by means of a special impeller or 'drift' so that the connecting nipple protrudes from the surface. With a standard grease-gun nozzle, attached to the spray lance of the normal spray equipment, the fungicide can be injected into the wood at fairly high pressure, forcing it along the grain until it emerges from the joint in the joinery. The vulnerable spots can thus be thoroughly treated with preservative which will destroy existing fungal growth and protect the wood from further infection. The inconspicuous protruding nozzle can be left or cut off flush with a sharp chisel. Putty and paint will then hide all signs of treatment. As the plastic injectors cost only a few pence each and the time taken to carry out the whole operation is only a few minutes, the process is quite inexpensive. The invention is that of a Frenchman, Marc Bidaux, but the injectors are marketed in Britain under the name Wykamol Timber Injectors.

Soft Rot

Soft-rot types of fungal decay generally occur in extremely damp places, like breweries, dairies and cooling towers, where the timber is very damp or wet for prolonged periods. Such fungi work from the surface inwards and leave a soft, mushy or pulpy skin on the outside of the wood, often without visible signs of mycelium or sporophores.

The timber should be tested to see if it is sufficiently strong for its purpose. Control treatment can be effectively applied by spray application of water-borne fungicides or organic-solvent type fungicidal preservatives. The aqueous forms of treatment should only be applied to timber containing at least 50 per cent w/w (weight/weight) but not more than 75 per cent w/w of water. This enables the aqueous solution to be absorbed and the dissolved salts to diffuse into the damp wood. To prevent subsequent leaching, these salts should be self-fixing to the wood or be precipitated in an insoluble form by a second treatment with precipitatory or fixing chemicals. For instance, copper sulphate solution can be suitably precipitated or fixed by means of sodium chromate or potassium dichromate solution.

In applying salts of this nature, the operator should be fully aware of the possible consequences of contamination of the water used in the plant or building which is being treated. He must be responsible for any measures necessary to eliminate all danger of contamination.

Organic solvent type preservatives should only be used to control soft rot after the affected wood has been dried to a moisture content of less than 20 per cent w/w oven dried.

Precautions against contamination and fire should be taken in accordance with the British Wood Preserving Association Codes of Practice.

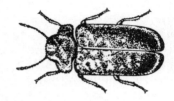

Fig 4 Death-watch Beetle (*Xestobium rufovillosum*) × 8

Treatment of Insects

Death-watch Beetles (Xestobium rufovillosum)

As Death-watch Beetle attack is most common in timbers affected by fungal decay, and this is brought about by dampness, care should be taken to ensure that the cause of the dampness has been remedied. All timbers should be probed, preferably using a stout, sharpened screwdriver. As has been observed, probing instruments which leave a circular hole often give, to subsequent examiners of the wood, the mistaken impression that the hole is caused by a boring insect. Large dimensioned beams and plates should be probed with auger bits to check the condition of the wood below the surface. Where there is external frass, this should be carefully and neatly removed by means of adze, axe, draw-knife or chisel and all the timbers carefully cleaned.

Type of preservative Organic solvent or emulsion-type preservatives with insecticidal and fungicidal properties should be used. The insecticides should produce their effect by contact and as stomach poisons and should be of a persistent character. The fungicidal action should be sufficient to destroy any type or species of fungus associated with Death-watch Beetle activity.

The preservative should contain a material to provide a prolonged surface deposit which would inhibit ovipositing by insects and spore germination of fungi, and should remain active for at least ten years.

Treatment of roof timbers When all timbers have been trimmed and probed and their strength assessed, any defective or weakened members should be clearly marked. Any hidden parts of beams, bonds, ties or plates, the ends of purlins, parapet and valley gutters, flat roofs and all hidden timber should be opened up and revealed, tested, trimmed and cleaned before treatment is applied.

Treatment of interior timbers Floors: all boards should be lifted and any other type of flooring removed to provide access to all affected sub-floor timbers. All such timbers should be carefully trimmed, cleaned, probed and/or drilled, and their strength assessed before the application of preservative treatment. Any weakness should be reported, and all necessary replacements should be carried out with timber pre-treated with an approved preservative, or thoroughly treated *in situ* when preservative treatment is applied to the remaining timber.

Structural timbers: studding and framing, sleepers and plates, bressummers, lintels, bonding timbers, wall posts and any timber buried or partly buried in the masonry, plaster, cement, concrete or soil should be opened up and revealed so that trimming, cleaning, assessment of strength, replacement (if necessary), and preservative treatment can be effectively carried out.

Joinery: panelling, linings and soffits of windows and doors, skirtings, wainscots, cornices and all such timber, which is affected or may be hiding affected timber, should be removed to enable effective treatment to be applied.

Staircases: all hidden timber should be revealed. Where strings (sloping board at each end of the treads, housed or cut to carry the treads and risers of a stair), against or partly buried in masonry, are affected, pressure injection is essential to enable the preservative to reach the hidden parts of the timber and fixings.

Method of Application As the Death-watch Beetle is most often found attacking old hardwood timbers of substantial size which are assembled by tenons and mortices and fixed with dowels, any treatment fluid should be applied so as to penetrate into all joints where it will reach the hidden tenons. This is best achieved by drilling and inserting Wykamol timber injectors, by means of which the preservative can be forced into the wood under high pressure. Active infestation deep inside the wood can also be dealt with by this means, although experience is needed both to locate such areas of injection and to ensure that the preservative reaches them. This method will enable the injected liquid to diffuse slowly through the vessels and grain to reach larvae boring inside the wood.

The timbers should then be sprayed with treatment fluid until all surfaces are saturated and the liquid runs off. The backs and undersides of wall-plates should be treated by high pressure spray and injection, where necessary. Bonding timbers should be drilled right through, fairly close to the top edge, at intervals of 6in, so that liquid can be injected at high pressure until it is seen to emerge from the sides, top and bottom of the bond. The top surface of large dimensioned tie beams, plates and camber beams should be drilled with 1in bits, staggering the holes at 9in intervals. Liquid should then be poured or injected into the holes to enable it to diffuse. Alternatively, an approved mayonnaise-type preservative may be applied in accordance with the maker's instructions.

Common Furniture Beetle(Anobium punctatum)

The following methods of applying remedial or control treatment also apply to hardwood infestation by *Ptilinus pectinicornis*.

Type of preservative   Only organic solvent or emulsion-type preservatives containing recognised persistent insecticides should be used. These should preferably be non-staining and approved by the British Wood Preserving Association.

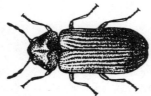

Fig 5 Furniture Beetle (*Anobium punctatum*) × 12

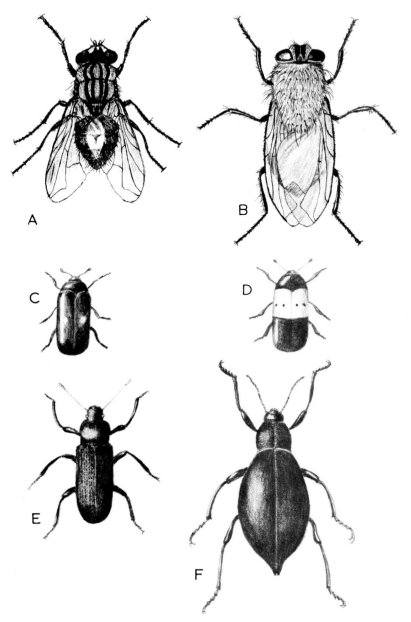

Plate 15 A House Fly *(Musca domestica)*; B Cluster Fly *(Pollenia rudis)*;
C Black Carpet Beetle *(Attagenus pellio)*; D Larder Beetle *(Dermestes lardarius)*;
E Mealworm *(Tenebrio molitor)*; F Churchyard Beetle *(Blaps mucronata)*

Plate 16 Simple injection treatment of furniture

Plate 17 Injection into Death-watch holes using a tapered nozzle

Treatment of roof timbers Seriously weakened members should be replaced, the badly riddled sapwood edges of hardwood timbers trimmed away and all the timber carefully cleaned by brushing and/or vacuum cleaning. Wall plates, parapet and valley gutters, apex cock-lofts and flat roofs are often difficult of access and should be opened up in such a way as to enable thorough cleaning and preservation treatment.

Treatment of interior timbers Floors: sufficient boards or other flooring should be removed to provide access for treatment to affected joists, plates, sleepers and any other hidden timber.

Joinery: infested panelling, the linings and soffits of windows and doors, skirtings and wainscots, cornices and any other ornamentations hiding structural or fixing timbers should be removed if a complete and thorough treatment is required.

Staircases: if the underside of the treads, risers and strings of infested wooden staircases are hidden by plaster or other ceiling materials, these should be removed.

Method of application In most instances, a sufficiently thorough and effective treatment can be given by spraying. The approved types of preservative are generally too mobile for application by brush.

It is rarely possible to apply an effective treatment to wood that is painted, varnished or otherwise surface-coated. Uncoated surfaces and flight holes which have penetrated the surface finish can, however, be successfully treated.

Ply or laminated wood is very difficult to deal with, and it is rarely possible to obtain penetration into the inner laminates, or to eradicate all insects with one application. If the plywood is uncoated, both faces should be injected through flight holes in the surfaces, then sprayed all over, paying particular attention to the edges. Coated plywood can only be effectively treated by painstaking and repeated injection through the flightholes.

Treatment of furniture This may require a different method of application of the insecticidal fluid or possibly an entirely different approach. Most articles of furniture are suitable for spray, brush and injection methods of applying suitable liquid insecticides, but some may be too delicate for this.

In such cases, fumigation is generally employed, but this should be carried out by experienced firms or individuals in special chambers designed for the purpose. Most effective fumigants are as dangerous to mammalian life as they are to the insects which they have been designed to destroy. Fumigation with a poisonous gas has the additional disadvantage of not leaving a persistent deposit of insecticide in the wood to protect it from further infection. Smoke fumigation is not satisfactory as the smoke, being a solid substance suspended in the air by thermal currents, will only settle on the surface of the article being fumigated and will not penetrate the flight holes, joints or fine crevices to which the insects can return and in which they may deposit their eggs.

Liquid containing a persistent insecticide in an odourless, non-staining, non-sticky and quickly volatile organic solvent is the most satisfactory material. It should never be sprayed or brushed over polished, varnished or painted surfaces. If insect holes are present in such surfaces, the liquid should be carefully injected using a syringe with a broadly tapered nozzle which has a hole at the tip no more than $^{1}/_{16}$in in diameter. Such a nozzle is self-sealing and will enable liquid to be injected at quite a high pressure. It will force its way along the galleries and tunnels made by the larvae, and so form a reservoir of insecticide which will permeate the wood.

The parts of a piece of furniture which are not polished or coated with varnish or paint are usually in the undersides, inside drawers and cupboards and generally hidden from view. It is necessary, therefore, to turn tables upside down, empty and remove all drawers, open up cupboards and turn them over, remove all cover boards and the key-boards from pianos. Everything necessary must be done to enable the treatment to be applied to those parts where the insects are most likely to deposit eggs and which are most receptive to the insecticide. Upholstered furniture must, in most cases, have the upholstery removed to apply liquid insecticide. The only alternative is to fumigate with a toxic gas. Panelling and similar forms of decoration should be removed if an effective treatment is to be applied.

Plywood is the most difficult material to treat, as the liquid cannot penetrate through the glue line, and the arrangement of the laminates prevents penetration. Unless the affected article is valuable, it is usually not worth the trouble to treat plywood and it is far better renewed or replaced with hardboard. Unfortunately, much modern furniture is made of plywood and compounded materials, and many owners will try and preserve it. Only injection into the flight holes and conscientious treatment of all uncoated surfaces by brush or spray, repeated each spring for about three years, is likely to succeed. Special attention should be paid to the edges of plywood, as it is in the joints and the exposed fibrous wood that the insects deposit their eggs. Plywood doors offer peculiar problems of their own which can only be overcome by removing the door from its frame, drilling with an $\frac{1}{8}$ in bit at 6in intervals all round the edges and through the framing timber. Insecticide fluid can then be injected through the holes and into the cavity spaces within the door. If the door is immediately rehung there is a possibility of the fluid leaking from the bottom edge. Some absorbent material, like sacking or corrugated paper, laid over polythene sheeting, should be put under the door to protect the floor from possible staining. Newspaper is quite satisfactory if thick enough layers are used.

Picture frames and even pictues, if on a wood backing, are often attacked by Furniture Beetles. Great care is needed to apply treatment and, if the picture is regarded as valuable, it should be carried out by experts.

House Longhorn Beetle (Hylotrupes bajulus)

Damage by this insect is confined to softwood timbers and is mostly restricted to the sapwood. Affected timber is not usually confined to structural members but may be found in joinery, picture mouldings, outbuildings, sheds and fences. As surprisingly few flight holes occur in relation to the amount of damage caused by the larvae of this insect, the presence of a single hole will certainly indicate that a considerable length of galleries exists below the surface of the wood. This should be uncovered by cutting away the surface. However, this often

Fig 6 House Longhorn Beetle (*Hylotrupes bajulus*) $\times$ 3

109

reveals several yards of galleries, some near the surface and some deep down, with a hard and unaffected layer of wood between. Thorough probing is necessary before one can be satisfied that the full extent of the damage has been discovered. When a number of flight holes are present in one piece of wood, it is rarely worth treating as it will, almost certainly, have lost most of its useful strength. In such cases, the affected members should be marked for removal and replaced with fresh timber suitably pre-treated by approved methods and with approved materials. Probing and the cutting away of affected wood releases considerable quantities of fine powdery frass which, together with the trimmings, should be removed, preferably by vacuum cleaning machine.

Type of preservative Organic solvent type with a persistent contact and stomach insecticidal action. It must be approved by the British Wood Preserving Association.

Method of application Pressure injection by means of special valve plugs inserted into the wood at 9in intervals. The holes must be staggered so as to ensure a flow of insecticide through all parts of the affected sapwood. This should be followed by thorough spray treatment as described for timber attacked by the Common Furniture Beetle.

Lyctus or Powder post beetles (*Lyctus brunneus*)

As this insect only attacks the sapwoods of hardwoods for a limited period, usually not exceeding ten years from the time of felling and then only if starch is present, it rarely causes damage of a serious structural nature in Britain. It is commonly found in stacked hardwoods in timber storage and seasoning yards, and in hardwood flooring, panelling and other ornamental timber and joinery.

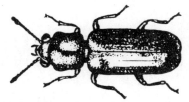

Fig 7 Lyctus Beetle (*Lyctus brunneus*) × 12

Treatment will be governed by the degree and extent of the damage, the accessibility of the affected wood, the finish on the surface of the wood and economic factors. If the wood is for ornamental purposes or flooring and is infested to the point where the affected parts are disintegrating, treatment is not justified. Such wood should be replaced with fresh wood which, together with all the wood in the vicinity which may be subject to infestation, should be thoroughly treated with a suitable, approved preservative. Wood, infested with *Lyctus* beetles but not damaged to the point of disintegration, can usually be effectively treated by a spray, using horticultural spray equipment, at low pressure and with a fairly coarse nozzle. The capillary effect of the small flight holes and borings is usually sufficient to convey the insecticide into the wood. Penetration can be assisted by sucking as much bore dust as possible out of the flight holes, using a vacuum machine, prior to applying the insecticide.

Treatment of block flooring is usually simple, for a generous application of organic-solvent type insecticide, similar to that used to control the Common Furniture Beetle, will liquefy the wax polish with which the blocks are usually coated and penetrate the affected sapwood through the top, sides and ends. As the practice is to lay the blocks on a pitch mastic base, treatment of the underface is unnecessary. Flooding the blocks, to the point where the joints are filled with insecticide, may cause the bitumen or pitch mastic to rise and stain the surface. This can be avoided by quickly removing any excess fluid with cotton waste.

Tongued and grooved strip flooring, invisibly nailed and tightly clamped, is a more difficult problem, but, unless the boarding is very thick — over an inch — one thorough treatment in May will usually be completely successful. This treatment should be applied to both the surface and underside wherever possible.

Affected panelling, furnishings and joinery which have been coated with french polish, cellulose or synthetic varnishes, paint or any other decorative finish should be taken apart so that the bare, unfinished surfaces are exposed. The insecticide should then be applied by brush or coarse spray over these exposed surfaces. Injection into the flight holes through the surface finish may be necessary where the affected sapwood does not continue through to the bare side.

Precautions

1   When roof spaces, lofts or other parts of a building where water tanks are situated are being treated with chemical solutions, gases or smokes for any purpose whatsoever, the tanks should be thoroughly protected with an impervious material such as polythene.

2   Where solutions are being sprayed in a situation such as a cellar or confined roof loft where the vapour could displace the atmosphere, adequate ventilation should be introduced to ensure a steady flow of air to evacuate the vapour and enable the operators to breathe comfortably. Breathing apparatus and respirators should be avoided unless the chemicals have or are likely to have any toxic effect. Such apparatus should be suitable for the particular purpose for which it is being used and should be of a type, quality and condition to provide adequate protection.

3   Fire prevention precautions should always be carried out in accordance with the Fire Code of Practice of the British Wood Preserving Association.

4   It is the responsibility of the firm using chemical solutions to satisfy itself that such solutions will have no harmful effect on the health of their employees. Written instructions should be issued to every operative on the safe handling and precautions that should be taken in handling and applying every type of chemical solution.

5   Washing, cleaning and sanitary arrangements should always be provided or arranged on every site where remedial preservative treatment is being carried out.

Chapter VIII

Timber Preservatives and Treatment Specialists

Very few owners of property appreciate the value of timber preservation, until their own is threatened with destruction by timber decay, and it is apparent that something must be done before very expensive structural repairs are necessary.

Unfortunately, treatment of *in situ* timber is often carried out by individuals and firms with no knowledge or experience of the complex problems involved, and with little or no sense of their great responsibility when dealing with valuable properties. They enter the industry armed only with a little knowledge, picked up from leaflets and advertising literature which proclaim panaceas for all timber troubles.

The production of a material for *in situ* treatment of timber is by no means simple. The evolution of a suitable material so often begins with a powerfully toxic substance which may not retain its toxicity sufficiently long to cover the life cycle of the insects, or cannot be satisfactorily conveyed into the wood by suitable vehicles, or is irritating, even dangerous, to the operators in confined spaces, or corrodes metals, softens veneers, or stains delicate woods, ceilings, decorations and fabrics. A material that could quite safely and successfully be used to preserve a fence, bridge, shed or garage may be quite unsuitable for treating the interior timbers of a dwelling house or church.

Materials for the *in situ* treatment of building timbers must fill a number of requirements. They must kill both fungi and insects and inhibit further infestation; in other words, they should be powerfully fungicidal and insecticidal, not only on application, but for a prolonged period afterwards. It is on this score that most, or probably all, gas fumigants fall short. Even the smoke fumigants probably do not have an active 'life' long

113

enough to embrace the life cycle of most wood-boring insects. Gas and smoke fumigation is a satisfactory form of treatment where it is required only to destroy existing insects and where reinfestation is unlikely to occur—for example, the destruction of tropical insects in timber brought to this country—but, for dealing with timber affected by indigenous insects, it is only likely to be effective for a very short time in the life of the building.

It is generally agreed that the organic-solvent type of preservative is the most suitable for *in situ* treatment of timber. Such a preservative must, of course, be insecticidal or fungicidal, or, preferably, both. As an insecticide, it should contain chemicals which act as stomach, respiratory and contact poisons and leave a deposit which will remain in and on the wood and retain its toxicity for many years. It must combine virulence with persistence thus removing the necessity for repetitive applications, if treatment, rather than replacement, is to make any sense. It is unlikely, of course, to kill all the larvae in the wood at once, but, in the course of time, these larvae will become adult beetles and will eventually emerge. If the wood is coated with a toxic chemical, then beetles which succeed in emerging will be destroyed. Other beetles which land on the wood, are unlikely to lay eggs on the toxic surface. If they do, the larvae will perish as they consume the poisonous deposit.

On the other hand, if the toxic chemicals are so volatile or unstable that they can only be applied successfully against insects on, or near, the surface of the wood, several applications must be given in consecutive years in order to exterminate each batch of insects as they emerge. This must continue as long as their longest possible life-cycle. Even if success is achieved and all the insects destroyed, no permanent deposit remains to prevent reinfestation, a possibility which may occur the year following the last of the repeated applications.

The solvent is another very important element. It must form a true solution with the toxic chemicals so that they are conveyed into the wood as far as the solvent penetrates. Very often, the solvent will penetrate but leave the toxic chemicals on the surface, and considerable experimentation is necessary before a solvent suitable to act as a vehicle for the toxic chemicals can be chosen. It should be volatile, but not too volatile. In such a case, the penetration is rarely satisfactory and during application such a vapour is produced that working conditions become impossible without suitable respirators—especially in confined roof spaces during the summer months.

Light and volatile petroleum solvents are generally favoured by manufacturers of organic solvent preservatives, but the lighter and more volatile they are, the greater the danger from fire by spontaneous ignition, electric spark, or some foolish act on the part of the workmen. Recently, there have been several fires in buildings where preservative treatment was being applied. One occurred because the electricity failed and the workmen decided to carry on with a candle, another where the electric cable insulation had been worn through by the passage of the workmen along a cat-walk in a church roof, yet another where a light bulb was dropped just as the atmosphere had reached saturation point, when the vapours exploded. This hazard has fortunately now been removed by the introduction of oil-based petroleum emulsions containing over 50 per cent water.

There are occasions, however, when infested timber has to be treated in cheese factories, wheat silos, oasthouses, tobacco warehouses, breweries and similar places where contamination of the contents is likely to occur. This calls for a completely odourless and non-poisonous solvent, which can still act as a penetrant and vehicle.

In most instances where *in situ* treatment is being applied, it is essential that the preservative does not stain timber, plaster, plasterboard or other linings. It should not soften, blister or change the colours of paints, varnishes, distempers or emulsion finishes. Nor must it corrode metals or cause irritation to the skin of the operators, harm their clothes or stain fabrics and upholstery. Finally, it must be reasonably cheap so that treatment costs are infinitely less than those for replacement.

Having satisfied the often demanding requirements for a suitable fluid material, one must find a method of application which will be effective against the offending species of insect or fungus. It is obvious that the methods used for dealing with *Lyctus* beetle, with its life cycle of only one or two years and its habit of attacking only the semi-seasoned sapwood of hardwoods, are unlikely to be so successful when employed against the Death-watch Beetle with its life-cycle of up to ten years and its predilection for fungus-infected wood, whether it be sapwood or heartwood, hardwood or softwood. The House Longhorn Beetle is another exception, for, although it confines itself mostly to the sapwood of softwood timbers, it packs its galleries so tightly with bore dust that the injection methods which might be successfully employed against the Death-watch Beetle are not so satisfactory against the House Longhorn. The Common Furniture Beetle is, of course, the most versatile of all the insect borers. It attacks hard and softwood, sapwood and, occasionally, heartwood, indigenous and tropical timbers, in pieces of any size from the finest veneers up to the largest structural timbers found in buildings. It seems to need no particular conditions of humidity and requires no help from fungi. Very often the greatest concentrations of Furniture Beetle flight holes occur on the polished, varnished or painted surfaces of the affected timber. Yet the eggs are never laid on such surfaces, but only on bare wood and, generally, in fine cracks, joints and crevices.

As well as these general truths, the operator must be aware of all the peculiar characteristics of the borers and fungal rots. Each man should be educated to the point where he can identify the common wood-boring insects by their flight holes, bore dust, larvae, pupae and imagos, and can also tell the difference between *Merulius lacrymans* and other types of fungal decay. Until he has this knowledge, he is in no position to apply the appropriate treatment.

Brush application with the light mobile liquids generally used for *in situ* treatment is quite out of the question. Those who advocate such a method have never attempted to treat overhead beams and rafters by dipping a brush into a very thin liquid and then trying to get it on to the wood. When the brush is lifted towards the timber, half the liquid runs down the handle, down the operator's arm and, very often, down his body into his boots. Various preventive devices have been designed for attachment to brush handles, but nothing less than a large umbrella will protect the worker for more than a few minutes.

Brush application with even the thicker and more oily forms of preservative cannot be wholly successful either. It can only be applied to the exposed and easily accessible faces of the timber, leaving the backs of rafters, purlins, wall-plates, the hidden sides of built-in bonds and studding timbers untreated. All organic solvents will penetrate just as well when properly sprayed on to wood as when brushed onto it, but the spraying must be controlled so that the surface of the wood does not become saturated too quickly and drip. A trained operator will move his spray lance backwards and forwards along a beam with such skill that he will gradually build up to saturation point without wasting preservative. This, of course, is very important when applying treatment to timber carrying valuable plaster ceilings. Even if the fluid is completely colourless, some staining always occurs if the preservative is permitted to penetrate, because the liquid picks up dirt and staining substances from the wood or plaster.

Although *Merulius lacrymans* is essentially a disease of timber, the treatment required to eradicate and control its growth is more far-reaching than the mere application of chemical preservatives to the timber. Timber preservatives intended to destroy it must be suitable for application to brickwork and masonry, concrete and other building materials. Certain chemicals which are powerful fungicides on timber lose their toxic powers through chemical reaction with lime, and there were many instances of dry rot recurrence before this fact was generally appreciated. Only a chemist could foresee such a reaction, and so it becomes a necessity for any responsible firm to employ a well-qualified chemist or to submit their formulae to chemists with a full knowledge of their intended use, the materials on which they are to be used and the conditions under which they are to be applied.

Oil-borne preservatives are obviously at a disadvantage when applied to walls or site soils saturated with water. The water-soluble preservative appropriate in these cases must be of a sufficient concentration to mix with the moisture of the walls or soil so that the consequent dilution will not fall below the effective toxic strength. It should also remain fixed in the treated material, and this can only be achieved by precipitation in an insoluble form in the material itself, or by the application of a second treatment with another precipitating chemical.

The old idea that brickwork, masonry, concrete or soil can be sterilized effectively by heating with a painter's or plumber's blow-lamp is absurd. All are such good insulators that it is quite impossible to heat them to the depth sufficient to destroy fungal hyphae and rhizomorphs which often penetrate for inches, and sometimes feet, into the fabric. Heating with flame-throwers which cover a fairly large area does assist in evaporating the moisture and raising the temperature to an appreciable extent, but such heating only has value in encouraging penetration of the fungicide and cannot be relied upon to destroy the fungal infection.

All this should be known to those who apply the treatment but it is on the technical surveyor that responsibility rests for the original diagnosis and assessment of the extent of the infection. Judging the extent of an outbreak of dry rot is probably the most difficult of all problems, and the ability to do so with reasonable accuracy can only come from long experience of observing outbreaks being opened up and fully revealed or, better still, by actual experience of opening up. Because experience is so indispensable, the training of technical surveyors should always include at least six months labouring under an experienced foreman.

Teaching workmen to appreciate the significance of the different forms of infestation and to apply the materials in the most efficient way is best done by visual aids. It takes the average man many months to memorize the names of the various species of insects and fungi he is likely to see in the course of his work but, with the aid of suitable films, he quickly learns to differentiate one from another and to know in which circumstances trimming, drilling, vacuum cleaning, injection, spray, heating, percolation, diffusion or complete removal will be necessary to cope with the situation.

These men will need more than a working knowledge of the 'enemy'. It is not work for those who have weak hearts, or who are subject to fainting fits. Indeed all workmen should be medically examined, and tested for their reactions to height. The tasks involved are strenuous and hazardous, often carried out at great heights on precarious ladders, or odd planks across tie beams or collars and in many other situations which call for good health and cool nerves.

When a firm specializing in *in situ* treatment is asked to deal with a building, their first move is to send a technical surveyor to examine the problem, diagnose the trouble, prepare a detailed report on the condition of the timber and any defects in the structure (inherent or otherwise), which might have given rise to the deterioration. They will suggest any structural modifications which might help to correct the existing situation or prevent further similar trouble, and then draw up a detailed specification of the methods and materials, together with costs, which must include labour, materials, hire of special plant (eg hire of generators where there is no mains current), scaffolding, transport, lodgings, insurances, overheads and profit. Such work requires a very wide knowledge of building structures and nomenclature, particularly when dealing with churches and ecclesiastical buildings and a head for heights even steadier than that of the workman for, quite often, only a precarious and worm-eaten ladder is provided for examining roofing timbers or steeples which will later be treated from scaffolding.

One of the surveyor's greatest problems is estimating the depth and extent of the frass which will have to be removed during the course of the work and how much work is involved in doing this. Because the frass must be safely removed and destroyed, the cost of this part of the work often accounts for a larger proportion of the labour charges than the actual application of chemicals.

He must also make certain that all the timber to be treated is accessible and that there is no timber hidden away, in which beetle and fungi can continue to flourish. This often creates the most difficult problems, the resolution of which calls for great moral courage on the part of the surveyor. It takes a lot of confidence to tell a client that all his sloping ceilings must be taken down because they hide the rafters and wall plates, or that his beautiful Adams cornice must be sacrificed to get at the dry rot growing in the fixing blocks or built-in bonding timbers. In so doing, the surveyor accepts a very great responsibility on behalf of his firm because all this kind of opening up and restoration is exceedingly expensive, and often costs far more than the preservative treatment itself.

For this and numerous other reasons, the framing of a report demands both ability and experience in order to arrive at the clear summary which both skirts legal pitfalls and fully satisfies professional responsibilities and ethics.

It is becoming an increasing practice of architects and surveyors to instruct timber decay specialists to examine the timber in houses which are about to change hands. They thus absolve themselves from any responsibility if insect infestation or fungal rot is subsequently discovered by the purchaser. There is a very great risk in this side of the specialist's business for, with all the knowledge in the world and using every device known to science, he could very well miss an incipent outbreak of *Merulius* which was completely hidden at the time of his inspection. This could become an extensive growth by the time all negotiations and legal formalities are completed and the client is in possession of the property.

Firms which specialize in church work find Christmas, Lent and Easter difficult periods, one bizarre consideration is whether the church is Roman Catholic or, if Church of England, whether it is high or low church. Roman Catholic and Anglo-catholic churches hold many more services than the low churches and, more often than not, insist on holding them even when the church timbers are being treated. This, of course, interrupts work every morning and occasionally at other times of the day if the incumbent is exceptionally devout or over-enthusiastic. Therefore, the charge for treating Roman Catholic and Anglo-catholic churches is appreciably higher than that for churches with less elaborate and frequent offices.

Dry Rot, Woodworm and Damp Treatment 'Guarantees'.

Firms engaged in the remedial treatment of insect infestation, fungal rot and damp in buildings usually issue some form of warranty or so-called guarantee which expresses the firm's own faith in the treatment it applies and which purports to protect the customers from financial liability, should the treatment fail.

Most 'woodworm' guarantees run for twenty years, it being argued that this is the period favoured by building societies, whose mortgages often run for that length of time. No other arguments have been put forward to explain this convention. So far as can be discovered, the first woodworm guarantee was issued in 1935. It was proposed when a new chemical formulation was suggested by a remedial treatment company for treating the roof timbers of Winchester Cathedral in an attempt to control the activities of the Death-watch Beetle.

As the formula was new and untried, the Dean and Chapter of the cathedral suggested it would be unreasonable to expect them to pay a substantial sum without any prospect of reparation. The company had confidence in its product and a guarantee was drawn up, based on the assumed maximum life cycle of the Death-watch Beetle, ten years. As the company was only just established, it was further agreed that, as the object was to rid the timber of the beetles, the guarantee should take the form of re-treatment until the beetles had been completely exterminated. Owing to the publicity given to the treatment of the cathedral, the terms of the guarantee were revealed and were then demanded by other authorities responsible for maintaining buildings.

After World War II, new firms entered the field of remedial treatment and, no doubt from reasons of commercial competition, issued guarantees covering longer periods. Fifteen, twenty, twenty-five and thirty years were written into the guarantees. Very few of the firms engaged in this class of work had any substantial financial backing but, in the scramble to become established during the years following the war, the sponsors and principals of firms with even a very small capital seemed to think the gamble against failure was worth taking. Hardly an advertisement offering woodworm treatment appeared without an enticingly generous guarantee. As more and more advertisements appeared, so architects, surveyors, and specifying authorities and the general public were taught to expect and demand a guarantee. As the majority of advertisers featured a twenty year guarantee, this soon became accepted as a standard, and only firms of long standing with unassailable reputations stood any chance of obtaining work without issuing a similar or even more generous guarantee.

The word 'guarantee' has crept gradually into ordinary use as a 'thing given as security'. In law, the term is usually applied to a contract by which a person called the guarantor or surety promises to answer for the payment of some debt or the performance of some duty, in the event of the default of the debtor or obligor, who is in the first instance liable. Thus three parties are involved; the contractor, the customer and the guarantor who acts as surety for the contractor. A man or a firm cannot be his or its own surety.

What cover does the client really have? If, as in most cases, the terms of the guarantee are only to re-treat timber showing signs of continued or fresh insect activity, then the guarantor firm can keep strictly to these terms, which may allow them to charge the client for transport and travelling, removal of floorboards or any other opening up that may be necessary. If, in an extreme case, this meant the removal of roof coverings such as lead or copper or even tiles and slates, or the stripping of plaster ceilings to reach hidden rafters, then the client might well be involved in vast expense in order to make it possible for the contractor to honour his guarantee, by carrying out the relatively minor operation of re-treatment.

Some of the larger remedial treatment firms are now including the cost of opening up and making good in the terms of their guarantees. This is, of course, of paramount importance where the guarantee covers the treatment of dry rot because, in general, the cost of stripping decorations, wall linings, fittings, pipings, and electrical wiring and then restoring everything to its original condition after sterilisation treatment has been applied, is often ten times greater than the actual cost of the chemicals and their application. Thus, guarantees which avoid mention of opening up and making good or specifically exclude them from the obligations of the guarantor, are offering very little to the client in the event of failure or unsatisfactory results. From our own experience, justified claims by clients against remedial treatment firms were invariably due to some failure of the human element, and not of the chemicals employed.

Many firms are now taking out insurance to cover their guarantees, but it normally only covers claims made during the years for which premiums have been paid. It is only in recent years that a firm of insurance brokers have been able to offer comprehensive insurance to remedial treatment firms. This cover is, however, experimental and might have to be reviewed or even dropped altogether if a series of substantial claims were made.

There is no doubt that most of the well-established and substantial firms who would have least difficulty in meeting any sort of claim, look forward to the day when membership of an effective Trade Association will, in itself, be sufficient warranty. There are instances of firms which carry out treatment with the

Plate 18 Drilling to treat dry rot in masonry

Plate 19 Simple pre-treatment of new timber by immersion

Plate 20 Pressure-spraying roof timbers

Plate 21 Trimming away riddled wood with a draw-knife and axe

products of one particular manufacturer forming a company, to which they all subscribe. The purpose of the company is to accumulate a fund in case any member firm is unable to meet its 'guarantee' obligations. Up to a point, this form of warranty represents a true guarantee but would have to underwrite every guarantee issued by these firms to have any real value. However, it is a step in the right direction and indicates that there are firms sufficiently ethical to try and overcome the shortcomings of the usual dry rot, woodworm and dampness 'guarantee'.

Should Building Timber Be Pre-treated?

In recent years much publicity has been given to the insects and fungi which attack building timber, and people often ask why all building timber is not treated with preservatives before installation.

Considerable sums of money are spent in stemming the ravages of timber-destroying organisms. Few would doubt that this expense and inconvenience could be avoided if the timber in the building had been effectively pre-treated at the time of erection. Most professionals concede that, in most houses or other buildings over thirty years old, particularly in rural districts, some form of timber deterioration is almost certain to be found, yet there are no statistics which make it possible to weigh average cost of pre-treatment against the projected average cost of remedial treatment.

Being aware of the hazards to which building timbers are exposed, many prospective purchasers of properties take the precaution of having them surveyed and carefully examined. Few however, consider these hazards when having a house built, or gauge their possible effect on the selling price twenty or thirty years later. Architect and builder will argue that damp-proof courses, cavity walls and adequate ventilation and drainage will completely safeguard the property from any form of timber decay. What happens, though, if the occupier builds up his garden or paths above the damp-proof course, the cavity walls become bridged with slovens, the air bricks become blocked and the gutters and hopper heads filled with leaves and debris? These are the usual causes of dry rot, and no designer or builder can ensure the proper maintenance of the house after it has been handed over to the owner. Beetles are quite unconcerned

125

with design and correct building practices. In summer, when windows are open, they fly vigorously from the dead trees, fences, old sheds and infected houses in the district and find very little difficulty in gaining access to the most perfectly designed and constructed building. The vast damage which can be caused by beetles is exemplified in north-west Surrey, where the House Longhorn Beetle has caused havoc, so much so that by-laws have been passed to check the spread of these voracious pests. A large proportion of the infested houses are owned and maintained by the local authorities, and the standard of maintainance could hardly be criticised. They provide graphic proof that good design, building and maintenance are no safeguards against insect infestation.

It may be argued that one by-law does not, and should not, an Act of Parliament make, and Surrey's harrowing experience should not force builders throughout Britain to use only pretreated timber. But the Furniture Beetle can be found in building timber from Land's End to John o'Groats and is, without doubt, a pest of considerable economic significance. It will attack almost any kind of timber used for normal building purposes — carcassing; flooring; joinery and decorations — but it usually confines its attention to sapwood. Unfortunately, owing to modern economics and methods of afforestation, the timber now used in buildings contains a far higher proportion of sapwood than would have been accepted before World War I. The obvious increase in the activity and scope of the Furniture Beetle is undoubtedly related to the inferior quality of timber used since that war, and most instances of very severe damage have occurred in houses built in the two decades after it. Today, many reports are coming in of Furniture Beetle activity in houses built since 1945 and it is not unreasonable to assume that the flow will steadily increase.

Investigation suggests that buildings erected during the last fifty years in large cities and in heavily industrialised districts, although equally subject to fungal decay, are not so susceptible to beetle infestation as those on the fringe of the cities and in rural areas. This is probably because much insect infection must originate in the dead trees of wooded districts, and also because atmospheric pollution has a controlling effect on insect activity and ovipositing. The differences between one and another make it preferable for regulations governing the treatment of building timber to be controlled by the local authorities.

Although the susceptibilities of timber in buildings are widely known and many scientific laboratories, government departments, ecclesiastical organisations and commercial bodies are concerned in various ways with fighting timber decay in buildings, there is no central organisation to collate their experience into a report. Because of this, no absolutely concrete case can be put forward to parliament or to municipal authorities for the pre-treatment of building timber.

The timber trade organisations have, in the past, appeared very reluctant to press for preservative treatment of timber. They are, apparently, afraid of over-emphasizing the dangers to which timber is prone, and thus of playing into the hands of those who are pressing for the use of concrete, steel, aluminium and plastics because they are not subject to decay. This seems very short-sighted thinking. It is because *untreated* wood has, on occasions, failed that substitute materials have gained favour. If treated timber had been used and was found to be efficient the advocates of the substitute materials would have had far greater difficulty in convincing architects and others that their materials offered any real advantages over wood.

Another factor which may have caused the timber trade to be reticent about preservation was the limited number of treatment plants available to deal with the processing if it became common practice. This problem has been largely overcome in recent years by the evolution of organic-solvent-type preservatives which give adequate protection and can be applied in the smallest of timber yards, or on the building site with the simplest of equipment. These new processes and materials can of course, be inadequately applied, and abused rather than used, but there is pressure to develop specifications for such processes, and thus make them no more subject to bad practices than the well-tried and universally accepted vacuum-pressure method of timber preservation.

Many whose business is timber decay believe that timber in all buildings designed to stand for more than thirty years, should be pre-treated with a preservative, the formula of which and the method of application being governed by accepted standards and subject to scrutiny and test by a recognised authority. A move in this direction has been taken by the National House-Building Council and cerain specifying authorities, such as the Greater London Council. The new Building Regulations will, almost certainly, cover the question of timber preservation and make recommendations, even if statutory regulations are not introduced.

A Cautionary Tale

It is extremely important to preserve a balance in our attitude to the insects and organisms whose destruction we have been discussing so single-mindedly. We must remember that each one of them has an indispensable function in the overall balance of nature. Wood-boring insects, for example, complement fungi in keeping forests clear of dead timber, which they reduce to valuable compost. Cockroaches, flies, wasps and all the other insects mentioned in this book perform similar tasks according to their eating habits and, in particular, the eating habits of their larvae. Fungi are probably the world's greatest scavengers and perform the most unpleasant of tasks, breaking down noisome substances to their elementary components, or creating the most complex of chemical reactions to change such substances to harmless gases. If all these confined their efforts to their natural environment, they might never be regarded as pests. It is only when they encroach on the province of that greatest of all pests *homo sapiens* that we see them in this light.

We may derive some comfort from the cases of others, whose plight is worse in this light than in our own. In Britain, one can state fairly safely that good construction, the use of carefully chosen materials and good maintenance and housekeeping will reduce to a minimum the possibility of such 'pests' damaging a building, or its contents. This cannot be so confidently said of buildings abroad, particularly where termites and longhorn beetles are active.

In such places, a perfectly constructed and well-kept house may become structurally unsound in a matter of weeks or even days without anyone being aware of what is happening. When I was in West Africa, my steward boy woke me one morning with the words. 'Dem bungalow he be no good, massa.'

I asked, 'What for you say he be no good, Gabriel?'

'Dem termite go chop um, Massa, in de lounge.'

'He chop um small-small or too big?'

'Him big too much, Massa, I tink you come one time and look-um.'

I went to 'look-um' and could see nothing wrong until Gabriel, to prove his point, kicked the end wall of the timber-built bungalow, which promptly collapsed in a cloud of dust. I then remembered that Gabriel had, several days before, pointed out the mud-covered galleries creeping up the walls of the bungalow and I had not then realised the dreadful significance of his warning that, 'him be no good for bungalow.' So, like most of us, I too learned the hard way.

Appendix I

Other Insect Pests

Spider beetles (Niptus hololeucus) (Ptinidae)

These beetles are often mistaken for tiny golden spiders but, as spiders have eight legs and beetles have six, the mistake should

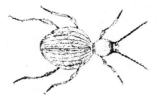

Fig 8 Golden Spider Beetle (*Niptus hololeucus*) × 10

be quickly realised. They often appear in very large numbers and cause considerable alarm to householders who fear they will damage fabrics, furnishings or woodwork. They are, however, only pests of dry stored foodstuffs.

Bacon Beetle (Dermestes lardarius)

Also known as the Larder Beetle, it is also a common pest of dried foodstuffs, such as bacon, preserved meats, fish and fish meal, cheese, dog food and other protein-rich foods. These insects can often be found in birds' nests and where food is consumed or stored by rodents. Being, very roughly, the size and shape of the Death-watch Beetle, they have caused alarm, particularly to occupiers of old houses in the country who are on the look-out for such pests. Actually, they are easily distinguished by a yellowish grey band across the centre of their backs. This takes up about half the space of their wing cases and highlights three or four tiny dark spots. The visible upper and lower parts of the beetle are almost black. The larva has the appearance of a small hairy caterpillar. As one female can lay up to 150 eggs it

130

is possible for an infestation to spread fairly rapidly, particularly in warehouses where hides or other animal matter is stored. In domestic houses where this beetle is discovered, a search should be made for a possible source of food, particularly in larders and under floors (for rodents' nests), as well as in roof spaces and birds' nests. Because grubs stop feeding when adult and move away from their food supply in order to pupate, they often burrow into any material which stands in their way. For this reason mature larvae, pupae or adults are often found in, and cause damage to, materials on which they do not feed. There are records of these insects in wood, books, woollens, cork, mortar, salt and sal ammoniac.

Churchyard Beetle (Blaps mucronata) and Mealworm (Tenebrio molitor)

These quite often appear in houses and, like the Spider Beetle and the Bacon Beetle, live on stored food products. They are both black, the former often mistakenly being called 'cockroach'. Both are relatively harmless, usually being found in stale flour, unclean bread-bins, under skirtings where flour or breadcrumbs have accumulated or where vegetable matter is decaying. The larvae of both are straight bodied, rather long and thin with a firm shiny body and clearly defined segments. The Churchyard Beetle is roughly 1in long and the Mealworm about $5/8$ in long.

Carpet Beetles — Anthrenus verbasci and Attagenus pellio

These are two beetles commonly found in buildings. Both are members of the Dermestidae family and, although they feed on all kinds of animal matter, including wool, fur and dried skins, both go by the name of carpet beetle. The damage they cause is often mistaken for that of clothes moth but the larva, which actually causes the damage, can easily be identified. Known as the 'woolly bears', the larva of the *Anthrenus verbasci* is only $1/4$ in long, brown and covered with hairy tufts, while those of the *Attagenus pellio* are about $1/2$ in long, also brown and covered with hairs but with a tail-like tuft. They are sometimes found wrapped up inside balls of woollen fibre or hair. The adult *Anthrenus* is a very small 'lady-bird' like beetle (only $1/8$ in long), broadly oval in shape with a dappled appearance,

Fig 9 Carpet Beetle (*Anthrenus verbasci*) × 10

produced by alternating patches of white and blackish or brownish scales. Its close relation, *Attagenus pellio*, is about a ¼in long, oval (but not so broadly oval as *Anthrenus pellio*), is black with a white spot in the centre of each wing case.

The adults emerge during the summer months and are able to fly quite vigorously. Attracted to light, they may spend the summer feeding on pollen and nectar. Damage in the house is, therefore, caused entirely by the larvae, and the holes which they cause are not associated with the excrement dust which occurs when clothes moths are responsible. A light spraying with a non-staining solution of Lindane or the sprinkling of insecticidal powder under carpets and into cracks and crevices in floorboards will usually effect a cure.

Clothes Moths

In the British Isles, there are only four small and very insignificant-looking moths which are officially labelled clothes moths and several others, equally insignificant, which are described as house moths.

The Common Clothes Moth (*Tineola bisselliella*) This is only about ⅓ in long, shiny golden-buff in colour and, when resting or running, appears narrow and elongated with the wings pressed closely to the sides of the body and with the edges together to form a ridge. The wing span is about ½in.

No damage is caused by the adult moths as their only purpose is to mate and deposit eggs on material on which the larvae might feed. Damage to clothing, furs, feathers, blankets and carpets is caused entirely by the voracious feeding of the larvae and it is these tiny maggots and pupae which one finds when examining the damage they have caused. The eggs are visible to the naked eye, they are only ¹/₂₄ in long, oval in shape and opalescent creamy-white in colour. They are usually found between the strands of woollen fabrics and at the base of the hairs of furs. Around the areas of damage, the moth larvae spin a web of silk and form cocoons. There is always a small cluster of excrement pellets mixed up with the silk and cocoons which will assist in distinguishing moth damage from that of carpet beetles.

Case-bearing Clothes Moth (Tinea pellionella) This is characterised by larvae occupying cocoons. The cocoons of the moth are covered with pieces of the material on which it is feeding, possibly as a camouflage and a protection for the soft-bodied larva or caterpillar. Only the head and fore-part, with three pairs of legs, protrude from the case. This permits the insect to feed and move through the material on which it is feeding. Other characteristics of this moth are similar to those of the Common Clothes Moth. They feed throughout the year in the larval stage, but winged adults are usually seen during warmish weather, chiefly in the early summer and autumn.

Large Pale Clothes-Moth (Tinea pallescentella) Despite the name, this is only about ⅔ in long with, at most, a 1in wing span. At one time a relatively rare insect, it is now found throughout England and Wales, but only rarely in Scotland. General habits are similar to the other clothes moths, but it is usually found in damper surroundings than the others.

White-Tip Clothes or Tapestry Moth (Trichophaga tapetzella) This variety of clothes moth is more commonly found in outbuildings than in houses. The winged adults are barely ½in long and are mottled black and white.

House Moths

Brown House or False Clothes Moth (Hofmannophila pseudos pretella) The adult female moth has a wing span of about ¾in and is about ⅔ in long, but the male is somewhat smaller. The wings are dark brown with blackish-brown spots. The larvae are about ¾in in length, white and rather more easily seen than those of the clothes moths, which are generally hidden under a silken web or in a cocoon. The larvae eats furs and skins, museum specimens of animals, birds, etc (amongst which it makes great havoc if it gains admittance to the cabinet drawers)' seeds, fruits and groceries of many kinds. The moth damages clothing, carpets and other woollen materials, herbs and drugs, paper and bookbindings, cork, horse-hair stuffing of furniture and a wide variety of other substances, which it consumes and uses for the construction of the cocoon in which it pupates. Possibly the most destructive of the 'clothes moths', the Brown House Moth has been found in many parts of the world ranging from Eastern Siberia to Australia.

White-shouldered House Moth (*Endrosis sarcitrella*)   This is also a brown moth, but smaller than the Brown House Moth and easily identified by the white head and thorax which gives it its name. It is often found floating on top of milk, although the larvae feed upon dried seeds, vegetable refuse, thatch, birds' nests, fungi and wine corks. It is regarded as being primarily a vegetable feeder but is recorded as having attacked carpets.

Prevention and Treatment of Clothes and House Moths

Both eggs and larvae are very susceptible to heat from sun, fires or radiators, although a temperature of 56°C is necessary for egg-albumen to coagulate. Ironing, also, will destroy eggs and larvae, and, in bygone days, it became the practice to carefully brush and hang on the clothes-line in sunshine all winter clothes before they were put away.

Most furriers provide cold storage facilities for valuable furs, and many dry-cleaning firms now offer a treatment for protection against moths as part of the cleaning process. Carpet manufacturers are applying moth-proofing chemicals during manufacture, such as DDT, Lindane, dieldrin and other chlorinated hydrocarbons. Regular and frequent brushing and cleaning of clothes, vacuum cleaning of carpets, 'sunning' of blankets and the use of anti-moth sprays and crystals applied in accordance with the makers' instructions will provide effective protection against moths.

House Ants

Thirty-six different kinds of ants are said to occur in Britain, and some of these appear in houses often enough to be called pests. Of these, the Small Black Ant, *Lasius niger,* is very common and may be seen in large numbers in sugar, on cakes and on most foodstuffs. They actually have their nests in the garden and may be found under anything which has been standing on one spot for some time.

× 2·5

Fig 10 Black Ants (*Lasius niger*) × 8

Dealing with these ants is not easy. They are so agile and are able to travel, relatively, such large distances over every conceivable obstacle, that no physical barrier can stop them. If every step, sill, air brick or other aperture in the house is generously sprayed with a solution containing chlornaphthalene wax and 'Grammexane', the ants will turn back from the treated spots and search for food in the garden instead of the house. Any attempt to discourage them by putting down poison bait is unlikely to succeed, for, as fast as they are destroyed, fresh ants will arrive. Prevention, in this case, is far better than attempting a cure. If a nest is located, a small quantity of the solution will generally be sufficient to destroy the whole colony, but one has to work rapidly, as ants very quickly evacuate disturbed nests. If, for instance, a flagstone is lifted and a large ants' nest is discovered, the stone should be replaced immediately, and the insecticide placed in an old tin or jam jar with a firmly fixed lid in which holes have been punched or drilled. The stone is then lifted and the contents quickly sprinkled over the nest. The above solution is obtainable under the trade name of 'Wykamol Plus' — it is actually designed for killing insects and fungi in wood, but is equally effective against ants.

Another common ant found in houses is the very tiny Pharaoh's Ant, which often nests in the walls or soil in the footings of buildings. These are also found on sweet foodstuffs, bread and biscuits, as well as meat, fats, butter and other proteinous materials. Only about $1/12$ in long, and reddish yellow, they keep to definite pathways, and if, for some reason, a path becomes unusable, the ants will start another elsewhere.

Poison bait is the remedy in this case, the idea being to get the worker ants to take minute quantities of the poison back to the nest and there give it to the young ants and possibly the queens as well. The technique of poison baiting is fairly simple. Small tin boxes (1½ × 1½ × 1in), or circular tin boxes (2in wide by 1in deep) are placed in the runs of the ants. They should be secured in position by an adhesive or nails, and two or three punctures should be made about ¼in from the bottom to allow the ants to enter and leave. The boxes should be numbered to ensure that they are easily accounted for and should be labelled 'Poison'.

Inside the boxes, the poison bait can be applied to blotting paper or cotton wool, but will probably be more effective in cake, liver paste or minced meat, as the base is attractive as well as the poison solution.

The formulae and preparation of poison solutions are as follows:

Percentage by weight

Water	51
Cane sugar	42
Honey	6.5
Thallium sulphate	0.5
Water	51
Cane sugar	42
Honey	6.2
Sodium fluoride	0.8
Water	51
Cane sugar	42
Honey	5.7
Sodium fluoride	0.8
Thallium sulphate	0.5

(Cane sugar and honey cane be replaced by an equivalent weight of syrup.)

Weigh out the ingredients, mix them, heat and stir until a homogeneous syrup is obtained. Do not boil and do not inhale any vapour given off during preparation. The poison baits are prepared by working carrier bait and poison bait solution into a wet paste.

Suitable quantities are:

Cake: 1cu in/poison-bait solution two teaspoonfuls (8cc).

Minced meat: 1cu in/poison-bait solution, two teaspoonfuls (8cc).

Finely chopped and rolled liver: 1cu in/poison-bait solution one teaspoonful (4cc).

These quantities are approximate, as they are liable to vary according to the carrier bait. Except under special conditions bringing about the early desiccation or the mould infection of poison baits, the latter will require renewal every three or four days. Experience will show what consistency best maintains the moisture and attractiveness of the poison bait.

A 0.25% Thallium sulphate bait cleansed a hospital of one of the House Ants in approximately three months after they had been a major pest for many years.

It should be remembered that *Thallium sulphate and Sodium fluoride are poisonous,* and care must be exercised in handling or using them. It may be well to employ a servicing firm specializing in this type of work.

Cockroaches

There can be few people who are not acquainted with the cockroach, although they may call it the 'black-beetle'. In fact, it is not a beetle at all but represents one of the most primitive forms of insect life, the Orthoptera. Cockroaches, 250 million years ago, were very similar to those we know today.

In Britain, only two species of cockroaches are commonly found in houses, the Common or Oriental Cockroach, *Blatta orientalis,* and the German Cockroach, *Blatta germanica,* but two other species may be found in warehouses, hot houses and on ships. They are the American Cockroach, *Periplaneta americana,* and the Australian Cockroach, *Periplaneta australasiae.* All these cockroaches are said to have been imported to Britain and appear to have been unknown before the sixteenth century. Even now, inspection of many kitchens and bakehouses,

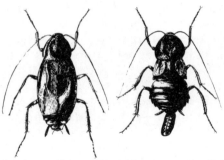

Fig 11 Cockroach — male (*Blatta orientalis linnaeus*)
Cockroach — female with oothecae or egg sac
Natural size

particularly in large establishments, would reveal the presence of cockroaches, sometimes in vast numbers.

They are most easily found at night when they emerge from their hiding places to feed on almost anything they can find. Their flattened bodies enable these loathsome insects to creep into very small cracks and crevices although, when bloated with food, they sometimes have difficulty in re-entering the refuge from which they emerged, and may then be seen scurrying along a wall looking for a chink. They gorge themselves when they emerge, and then return to their hiding places where they stay for several days whilst the food is digested. Only a small proportion of a cockroach population is likely to be seen at one time.

The sexes of the Common Cockroach differ markedly from each other. In the male, the wing-cases, or elytra, and the wings are more strongly developed than in the female. The general colour is not black but very dark brown. The antennae are nearly as long as the body and composed of over 100 segments. They taper away to form whip-like appendages like a couple of miniature fishing rods.

In both the common species, females greatly outnumber males and pairing takes place as soon as sexual maturity is reached. The eggs are contained in a purse-like pouch which forms within the abdomen of the female. This emerges after about five days when it remains attached to the body of the insect for up to another fortnight, after which she will generally deposit it in a warm, sheltered spot near a supply of food. The incubation period appears to vary inversely with the temperature in which it is kept. About thirty larvae or nymphs hatch out and can develop, through about seven moults, to maturity in six weeks or so, depending upon temperature and availability of food. This will indicate how rapidly they can multiply.

The control of cockroaches has become a very serious study, and many toxic chemicals have been employed both experimentally and in the practical field for achieving eradication. The chlorinated hydrocarbons such as DDT, Lindane and dieldrin are very successful for a time, until the insects build up a resistance to the insecticidal effect of these chemicals. Specialist firms usually vary the choice of insecticide periodically, to overcome the build-up of these resistant strains. It is therefore advisable to employ an experienced and knowledgeable organisation or individual to apply the appropriate and effective treatment.

The simplest way for householders themselves to deal with cockroaches is to use Gammexane dust (ICI), and, using a horticultural insecticide dust blower, to blow it under skirtings, behind cooking stoves, in cupboards and wherever cockroaches are suspected of hiding. If the whole colony is destroyed no resistant strains can be produced, as this only occurs in the second generation of insects which have survived chemical treatment, subsequently mating and producing offspring which are inured to the effects of the insecticide.

Proprietary insecticidal baits are also available as an alternative to sprays and dusts which are safer to use in places where there are animals or live fish and where there is a risk of food contamination. The baits are supplied in a form ready to use and often in containers which may be placed in areas to which cockroaches have easy access but other animals cannot reach. They may be in the form of cereal pellets, in sugar/starch mixtures, pastes or gels. The chemical most favoured as an insecticide for the purpose is chlordecone, a chlorinated ketone sold under the trade name Kepone.

Earwigs

Earwigs are one of the most maligned of all insects. Actually they are completely harmless, and there is no foundation to the

Fig 12 Earwig (*Forficula auricularia*) × 2

story that they favour the interior of the human ear. The formidable-looking forceps at the end of their bodies can cause no harm, and are used in unfolding the insect's complicated wing structure.

Their natural habitat is in the fields and gardens, where they remain hidden during the day and emerge at night to feed on other insects, fallen fruit, the sap of flowers and other plant matter. They are often brought into the house in flowers, fruit and vegetables. It is advisable therefore to inspect all such produce in the garden. A gentle shaking will usually dislodge the earwigs. Do not destroy them, but let them continue their useful work of consuming aphids and other harmful garden pests and decaying vegetable matter.

When earwigs are found in a house it is usually under damp material or in the folds of curtains. They rarely occur in sufficient numbers to justify a serious campaign of eradication, but, if they do become intolerable, a dusting of Gammexane dust under the mats and on the curtains will destroy them and discourage others from hiding in the same spots.

Flies

There are, of course, many different fly species and quite a number of these may enter the house at some time or another.

Common Fly or House Fly (*Musca domestica*) This is about 10mm long, grey, with four black stripes running vertically up the thorax. The abdomen is lighter and almost buff in colour; the wings are transparent and clearly veined. The eyes are conspicuous, particularly in the male. This description might fit other species of fly and, in fact, it is not easy to identify this one from other house visitors.

Horse manure, pigs' dung, human faeces and decaying matter of almost any kind will serve as a breeding place for the

House Fly provided it is moist, not too cold and can be easily consumed by its maggots or larvae. The eggs are laid in such materials and the maggots emerge in from eight to forty-eight hours. The eggs are glistening white, long-oval in shape, about $\frac{1}{20}$in long and are usually laid in small masses, which makes it easier to locate them. They are always laid in the material on which the maggots can feed and the females will often cluster in large numbers on a piece of such material, all engaged in laying their quota of 120 to 150 eggs at one time. As each female can lay 120-150 eggs five or six times, she may lay as many as 900 before dying.

The maggots feed vigorously, grow rapidly and can reach full size in three to four days under the most favourable conditions. In cool weather, the period taken from the hatching of the egg to the formation of the pupa may be as long as two months. In the pupal stage, the insects can survive a whole winter and spring. When fully grown and ready to pupate, the maggot is slightly under $\frac{1}{2}$in long, creamish and waxy in appearance. It rarely pupates on the rotting material on which it was feeding, but usually finds a cool and sheltered spot in which to develop the pupal casing which, when first formed, is yellow but gradually turns to red, brown and black.

Flies can infect food in three ways. The insect, in feeding, liquefies its food by vomiting upon it the remains of its last feed and then sucking back the resulting mess. During feeding, the fly often passes a drop of liquid excrement which falls upon the food and lastly the fly walks all over the food with faecal contamination on its feet. In this way, flies may spread typhoid, dysentry and diarrhoea, especially summer diarrhoea in children. In fact, it can also transmit many diseases caused by bacteria, viruses or parasites (such as intestinal worms), and must be regarded as one of man's most serious pests.

There are many reliable products available in large or small quantities designed to kill flies and prevent them from breeding. The most convenient for the ordinary householder are those which emit a vapour harmless to mammalian life but highly toxic to insects, in the form of aerosal sprays. The use of DDT, Dieldrin, BHC and other persistent chlorinated hydrocarbons has been condemned in recent years, but pyrethrum derivatives and other harmless insecticides have been developed which are extremely effective. There are specialist firms who provide a regular fly-proofing service which might, with advantage, be employed in large establishments. Farmyards, skin merchants and tanners' establishments, rubbish dumps and the like, act as the principal breeding grounds, but measures are being taken, sometimes on a very large scale, to deal with these sources. The ordinary householder can help considerably by keeping refuse bins clean and sprayed weekly with suitable insecticide. Regular and frequent burning, where possible, of refuse in an incinerator is a very effective method of keeping down the fly population.

Lesser House Fly (Fania canicularis)   Almost as dangerous as the house fly, it is capable of contaminating human food, being a frequenter of rubbish heaps and the soil of chicken runs, in both of which it breeds in a manner similar to the House Fly. The larvae, however, are different from those of the House Fly, being set with bristles and spines, as also are the pupae.

Fortunately most people think that the House Fly and the Lesser House Fly are one and the same insect, and so deal with them in a similar manner. This is to be encouraged, and, indeed, the destruction of all flies is excusable since most of those seen in domestic dwellings in Britain are unpleasant creatures likely to cause harm. Thus the Blue-bottle, Green-bottle, Biting Horse Fly or Stable Fly, Flesh Fly and Small Fruit Fly can be summarily dealt with by aerosol spray or fly-swatter.

Cluster Fly (Pollenia rudis)   Although these flies, being larger than House Flies, have a more repulsive appearance than the other flies one is likely to see in the house, they are in fact almost entirely harmless and unlikely to cause any contamination of food or transmission of disease. The Cluster Fly lays its eggs in the earth during September and the young larva then enters the body of the earthworm (*genus Allolobophora*), where it remains through the winter. In the spring, it becomes active and gradually consumes the interior of its host until little remains. In this it pupates and the adult fly emerges after about thirty-five to forty-five days.

The wings of the adult fly are closed, one over the other, like the blades of a pair of scissors. They enter houses in the late summer or early autumn and congregate in groups, in the dark apexes of roof spaces, behind shutters and in other dark places. Quite often they remain undiscovered until they fall away from the cluster, usually on warm days in the late winter or spring, and make their way to windows. There they lie or stand in a half-comatose state, some on their backs slowly waving their legs, and others crawling lethargically just an inch or two at a time.

The usual aerosol spray, together with a Gammexane smoke cannister, ignited in the roof space, will generally produce a cure.

Wasps

Wasps, contrary to common belief, perform a useful function as predators of other more serious pests. They feed their larvae mostly on other insects, the majority being flies. Butterflies, moths, bees and spiders are also caught and fed to the larvae, though in much smaller numbers.

In Britain, there are a considerable number of insects which, entomologically, are described as wasps, but laymen are likely to confine this description to the six species of social wasps and the hornet. By far the most common and most likely to be seen inside houses are the two ground-dwelling wasps, *Paravespula vulgaris* and *Paravespula germanica*. They generally build in the ground, in earth banks, river banks and subterranean sites, but occasionally choose roof spaces, wall cavities, shutter cases and even unused chimneys as nest sites.

Hibernating queens may often be found in the folds of curtains, stored clothes and blankets and, when disturbed, can inflict a vicious and painful sting. The workers and males all perish in the late autumn, so that an entirely new colony has to be established each year by one or more of the surviving queens. All these forms of wasp are capable of inflicting stings, but usually only in self defence, or when suddenly disturbed.

The natural function of wasps is to act as scavengers, and they serve an extremely useful purpose in consuming vast quantities of over-ripe and rotting fruit. This is actually eaten by the adult wasps, the young larvae, as previously mentioned, being carnivorous. This carnivorous habit means that adults will alight on meat and fish and this justifies their being regarded as pests to be destroyed.

The aerosol fly spray already suggested is quite effective against visiting wasps, and the puffing of Gammexane smoke into the nest will destroy a whole colony. This should be carried out after dark when all the wasps are likely to be in the nest and are less likely to emerge and sting their attacker.

Hornets

The one common hornet in Britain is *Vespa crabro*. It is appreciably larger than the common wasps, has a black breast and a yellow rear part with a dark brown pattern.

It is a terrible predator, hunting down other insects not only for its own food but also to chew and regurgitate for its larvae. It can be a very serious menace to honeybees and causes beekeepers considerable anxiety.

Hornet stings are painful, but rarely dangerous unless some particular part of the body, such as the tongue, eyelid, nostrils or in fact any part which could be seriously constricted by the swelling, is affected. My own particular remedy for wasps', hornets' or bees' stings is to dab Sal volatile on the area affected. The spirit has a cooling effect and the ammonia neutralizes the acid. There are records of human and other mammalian creatures being killed by stings inflicted by wasps or hornets. Usually, many stings have been inflicted on the victim, but there are instances of allergy where only one sting will cause death. This allergy is usually produced by a previous sting or stings which renders the person hypersensitive and deaths are extremely rare.

If a hornets' nest is found, a small heap of Lindane dust spread over the entrance will generally lead to the destruction of the whole colony. As the wasps leave or return to the nests, they will pick up some of this highly toxic insecticide. Only a minute amount of this chemical proves fatal, and so every wasp or hornet in the nest will die. If the nest is fairly accessible, a 1 per cent solution of Lindane in kerosene or white spirit, applied at night, will immediately destroy everything in the nest, which may be removed and burned in the morning. A horticultural spray with a fairly long lance should be used. The nozzle should be thrust through the casing surrounding the nest and the insecticide sprayed right into the centre, twisting the nozzle about until the interior is saturated. The whole outside of the nest should then be sprayed. If carried out neatly and quickly, the chance of being attacked by the wasps or hornets is very remote.

Silver Fish, Lepisma Saccharina and Fire Brat (Thermobia domestica)

These are the odd-looking insects which scurry around drawers, larder shelves and fireplaces. They are wingless, with long antennae in proportion to their bodies, and with peculiar, spiky, tail-like appendages. Both act as scavengers, consuming starchy foods which they will sometimes obtain from stored clothing.

In the days when men's shirt fronts and cuffs were stiffened with starch, Silver Fish would often consume little wriggly channels which led to the laundries receiving complaints of damage and negligence. The Fire Brat is almost unique in that it can withstand high temperatures, and may often be found exploring hot ovens. They are more often seen at night than during the day and, although they may be revealed if a drawer is opened quickly, they will soon move rapidly away. They are usually about ½ in long, and sharp eyes are needed to spot them. I have come across these insects in every part of the world I have visited from Norway to Australia and particularly in Africa.

Any proprietary moth or fly spray will effectively destroy Silver Fish or Fire Brats but be careful to read the directions and carefully observe the precautions printed on the containers.

Book Lice

These are probably the smallest insects likely to be seen in buildings and, as their main diet is fungal mould, they are associated with dampness. Stale, starchy materials such as bread, flour and damp paper serve as breeding grounds for these very tiny insects. They will often be found in a damp cold atmosphere, amongst books and papers which have been undisturbed for a long time.

For this reason, they are associated with libraries and book cases. They do not themselves cause any real damage, as a rule. The fungi on which they feed is the real danger, but they do serve to indicate that a high degree of dampness is present which may be due to direct moisture penetration or to condensation.

If the condition is effectively dealt with, the fungi will become inactive and there will be no further deterioration. The insects can be ignored, but, if it is felt that they should be eradicated, the use of a special material called Bibliotex is a good idea. This is a very volatile liquid, containing a fungicide and insecticide, which is applied by means of a very fine spray. It will cause no harm to paper, print illuminations or anything likely to be found as part of a library. A special grade of Bibliotex is also prepared for treating leather and parchment.

Appendix II

Deterioration in Libraries and Museums

In describing some of the insects and fungi, I have made brief reference to their effect on books, works of art and museum materials. Many varieties of insects and fungi can cause damage, in one way or another, to the contents of libraries, museums and picture galleries, sometimes with catastrophic effect. Books and their bindings may have four or five different species of insect and several varieties of fungal mould attacking them at the same time. Both insects and fungi are attracted to books only when they are stored in a damp atmosphere and it is, in fact, the decay produced by fungi which causes the most damage and attracts the insects which may be found in the paper, binding, board, leather, parchment and glue of which they are composed.

Different insects will attack different parts of a book, some attacking the cellulose and starch or the vegetable components whilst other attack the protein or animal constituents. Book Lice, Silver Fish, Carpet Beetles and Furniture Beetles have already been described but another relevant member of the same family as the Furniture Beetle, is the Drug Store Beetle (*Stegobium paniceum*). This beetle is so named because it was often found in the drawers and containers in which the pharmacists of old stored their root drugs. particularly orris root, ginger, turmeric and similar herbal drugs. Today, it might well be named the 'Dog Biscuit Beetle', for it is greatly attracted to stale or damp dog biscuits in which it will breed very rapidly.

This insect, together with *Anobium punctatum,* will often be found attacking the boards of larger books and often penetrates into the pages, perforating them through and through so that one can actually peep through a 'gallery' from one cover to another.

Leather bindings, parchment, vellum and other materials of animal origin are consumed by the *Dermestidae* (Bacon Beetle, Carpet Beetles and Hide or Leather Beetle). The Hide Beetle, *Dermestes maculatus* or *vulpinus,* is found in libraries and museums all over the world, feeding on dead animal matter, leather, taxidermy specimens, horn, feathers, hair, entomological collections, indeed all the trappings of museums and libraries.

Clothes Moths and House Moths can also cause serious damage, not only to fabrics, but to many other stored products and, in particular, collections of mammal and bird skins, insects and other animal matter.

The damage to water colours, prints, etchings and oil paintings is caused by similar fungi and insects to those which attack books but the most serious is generally caused by the Common Furniture Beetle. This infests the frames and backing and extends larval galleries and flight holes to the pictures themselves.

All these pests can be easily and effectively dealt with by using Bibliotex No 1 on vegetable matter and Bibliotex No 2 on animal matter *but it should never be used on oil paintings*. If the frames are treated, they should be allowed to dry in a well ventilated and warm atmosphere for twenty-four hours.

Modern insecticides and fungicides have been compounded in harmless media to deal with such pests and the discovery of the long-lasting effects of tri-n-butyltin oxide and hexachlorcyclohexane has provided the custodians of libraries and museums with extremely efficient means of eradicating and preventing pests of the above character.

Appendix III

After the Flood

The immediate effect of burst pipes and flooding, whatever the cause, is obvious and most distressing. But, provided the householder has been sensible, the expense of restoration and making good is usually covered by insurance. Unfortunately, though, there are more insidious and less obvious dangers, which may give rise to greater and more expensive damage, and these are seldom covered by the most 'comprehensive' of comprehensive policies. The danger arises from the germination of fungi during a period of dampness. The growth can flourish and extend long after all appearance of dampness has disappeared.

Such a fungus is dry rot, fully covered in my chapter on fungi but it cannot be emphasized too strongly that dry rot always requires damp conditions for the germination of spores. The spores require several weeks of dampness before they begin to germinate, and if this fact were generally realised, many outbreaks of dry rot could be prevented and millions of pounds saved.

The only real answer to the potential menace of dry rot is to *dry everything quickly*. This may be easier said than done at times, but a combination of ventilation and warmth is required. In centrally heated houses it is only necessary as a rule to open up all areas to which the damp has penetrated, so that evaporation can take place without leaving behind stagnant pockets of moisture-laden air. 'Opening up' means removing floor coverings and floorboards at regular intervals, as well as panelling and anything else behind which there is likely to be stagnant air. All cupboards, under-stair spaces, spandrel spaces over eaves and similar unventilated areas should be opened and should remain open for several weeks, or until the areas are absolutely dry.

In houses warmed by localised heating, the problem is more difficult. Where electricity is available, the blower-type heater is the most satisfactory appliance for heating and ventilating because the warm air can be directed to the exact area, and the heater should never produce sufficient heat to risk a danger from fire. Houses often have been set on fire by careless use of heaters drying-out after pipes have burst. Ordinary electric fires should be used at a safe distance from ignitable materials and convector heaters should never be turned on their sides or upside down to aim heat in the right direction. This misuse of appliances has, again, caused serious fires after flooding. A good emergency substitute for an electric fan is the 'blowing end' of a vacuum cleaner with the bag removed. Portable gas fires and oil heaters generate moisture and fumes and should only be used with caution and in well ventilated areas.

If there is any possibility that dampness has not been entirely eliminated, the area should be sprayed with a suitable fungicide.

Bibliography

Bletchley, J. D. *The Biological Work of the Forest Products Research Laboratory, Princes Risborough iii. The Work of the Entomology Section, with particular reference to the Common Furniture Beetle* (Proc Linn Soc 168, 1957, 111-15)

Cartwright, K. St G. and Findlay, W. P. K. *Decay of Timber and its Prevention* (HMSO, 1959)

Cornwell, P. B. *Pest Control in Buildings — A Guide to the Meaning of Terms* (Hutchinson, 1972)

——. *The Cockroach* 2 vol (Hutchinson, 1968)

Findlay, W. P. K. *Dry Rot and other Timber Troubles* (Hutchinson, 1963)

Hickin, Norman E. *The Insect Factor in Wood Decay* (Hutchinson, 1963)

Laing, Frederick 'The Cockroach', *British Museum Economic Series* No 12 (BM, 1930)

Richardson, B. A. 'Control of Biological Growths' (*Stone Industries* Mar/Apl 1973)

Richardson, S. A. 'Biological Deterioration of Ancient Buildings' (*British Wood Preserving Association News Sheet* No 113, 1970)

——. *Surveyor's Guide to Timber Decay, Diagnosis and Treatment* (Wykamol Ltd, Winchester, 1972)

——. *The Legal Responsibilities of Timber Infestation Surveyors and their Employers* (British Wood Preserving Association Annual Convention, Cambridge, 1975)

151

Acknowledgements

So much of the experience and knowledge I have incorporated in this book was gathered from biologists, chemists, architects and others who acted as preceptors and gave me encouragement but now remain only as memories.

I feel I must pay particular tribute to the late: T. D. Atkinson, Architect to Winchester Cathedral; A. P. Lay, Official Architect to the Church Commissioners; Harold S. Sawyer, Diocesan Surveyor, Winchester Diocese; W. J. Donger, Diocesan Surveyor, Winchester Diocese; Dr Ronald C. Fisher, Forest Products Research Laboratories; Colonel Guy Elwyss, aesthete; Bertram J. Starling, fellow director and a petroleum engineer; W. Carpenter (Billy) my first foreman and expert practical builder, and, for cajoling, bullying, checking, criticising, arranging, typing and making the book possible, my grateful thanks go to Mary Wheeler, my very live secretary and an author in her own right.

All photographs are by courtesy of Wykamol Ltd and the drawings are the property of the author.

THE AUTHOR:

STANLEY RICHARDSON is founder and chairman of Richardson & Starling Ltd and VIV Pharmaceuticals Ltd. A pharmaceutical chemist and a Fellow of the Institute of Wood Science, he was called in to help preserve the roof timbers of Winchester Cathedral, and evolved suitable methods and the product — later known as 'Wykamol' — on which his company was originally based. His expertise, humour and appreciation of the ways of building owners or curators, as well as the ways of pests and fungi, are unique.